Classroom Companion: Economics

The Classroom Companion series in Economics includes undergraduate and graduate textbooks alike. It welcomes fundamental textbooks aimed at introducing students to the core concepts, empirical methods, theories and tools of the field, as well as advanced textbooks written for students at the Master and PhD level seeking a deeper understanding of economic theory, mathematical tools and quantitative methods.

Gerald Shively

A Beginner's Guide
to Dynamic Optimization
in Economics

 Springer

Gerald Shively (iD)
Department of Agricultural Economics
Purdue University
West Lafayette, IN, USA

ISSN 2662-2882   ISSN 2662-2890   (electronic)
Classroom Companion: Economics
ISBN 978-3-032-09373-8   ISBN 978-3-032-09374-5   (eBook)
https://doi.org/10.1007/978-3-032-09374-5

This Springer imprint is published by the registered company Springer Nature Switzerland AG.
The registered company address is: Gewerbestrasse 11, 6330 Cham, Switzerland

If disposing of this product, please recycle the paper.

For Ben and Katie

Preface

Static optimization and dynamic optimization are important tools in economic analysis. In static optimization, one uses mathematical methods to find the maxima and minima which result from optimal decisions or choices, subject to constraints. In dynamic optimization, we similarly seek to find maxima or minima, but in this case from a series of decisions along a path. In economic problems, this path typically is indexed by time. Rather than seeking a solution point, as in a static optimization problem, we seek a complete set of solution points extending from the beginning to the end of the problem. This set of points constitutes an optimal path.

The main tools of dynamic optimization are variational calculus, optimal control theory, and dynamic programming. Although these approaches overlap somewhat, each has particular advantages and disadvantages and appropriate spheres of application within economics. For beginners, dynamic optimization is frequently intimidating because it is easy to feel overwhelmed and to get mired in unfamiliar concepts and new mathematical details—especially proofs. My aim in writing this book has been to approach dynamic optimization in the most basic and intuitive way possible, gradually adding key concepts and small doses of mathematical development as needed. The treatment is as light as possible. My hope is that this approach will help to engage my readers while at the same time establishing clear intuition to preempt confusion and anxiety. Armed with basic intuition, interested readers should then be in a position to access material offering greater levels of detail and rigor.

I developed many of the ideas in this book while teaching dynamic optimization to second-year Ph.D. students in applied economics. My target audience includes such students, many of whom might be comfortable with advanced microeconomic theory and static optimization but who are encountering dynamic optimization for the first time. In such a case, this book might serve as a primer to standard treatments, including Chiang's *Elements of Dynamic Optimization* or Kamien and Schwartz's *Dynamic Optimization*. My audience also includes anyone who would like some initial exposure to the subject. This includes advanced undergraduates, scholars from other disciplines, and anyone else interested in understanding how economists think about problems that unfold over time.

As for the somewhat tangential footnotes and references sprinkled throughout the text, I can only invoke in my defense that great character of the imagination, Sherlock Holmes, who in the story *The Valley of Fear* suggests that "the interplay of ideas and the oblique uses of knowledge are often of extraordinary interest." I trust my readers will agree.

West Lafayette, Indiana   Gerald Shively
July 2025

Competing Interests The author has no competing interests to declare that are relevant to the content of this manuscript.

Contents

Basics

Nearly everyone likes a party, and every party needs a cake, so let's begin with that.

Eating Cake

Imagine you are hosting a party and providing a cake for your guests. The cake looks delicious, and your guests are eager to dig in. Your job is to slice the cake and distribute the slices, making sure everyone has an opportunity to partake. At this point, we could complicate the story a bit, say by adding ice cream or specifying that our guests have specific likes and dislikes. We could even put some of the guests on a diet. "Oh, only a thin slice for me, please." Instead, we will keep it simple for the moment. Indiscriminate and gluttonous guests.

OK. The cake is on the table with forks at their ready. And there are plenty of forks because, of course, you are very popular. This raises a potential complication. If the total amount of appetite among those to be served exceeds the available cake, then choosing the size of each slice becomes a problem you must solve. To put it in economic terms, we are trying to satisfy everyone's pleasure from consuming (utility, in econ-speak), subject to a constraint on available resources. Pleasure comes from eating cake. But our collective pleasure is constrained by the size of the cake. How should we proceed?

Optimization subject to constraint is the central building block of economics, and it lies at the heart of this problem. Not surprisingly, to make progress even in this very simple example, we need to make some assumptions. First, we need to assume everyone wants a bit more cake than they are likely to get.[1] Second, we

---

[1] In practice, we might not know exactly how our guests feel about cake. Incorporating individual preferences adds a wrinkle to the problem that we cannot easily iron out. For now, we can make

© The Author(s), under exclusive license to Springer Nature Switzerland AG 2025
G. Shively, *A Beginner's Guide to Dynamic Optimization in Economics*, Classroom Companion: Economics,
https://doi.org/10.1007/978-3-032-09374-5_1

need to assume our cake is the only one available. In other words, we don't have the option of running to the corner bakery to pick up another cake.

Everyone has an intuitive approach to solving this hypothetical allocation problem because many of us have occasionally had to solve an actual and practical version of it. In the absence of any additional information, we tend to approach it with fairness in mind, choosing equally-sized slices of cake for everyone. That tends to work well in practice. But if you think more deeply about the problem, you will find that efficiency concerns also find their way into our intuitive approach. For example, if the cake is small, we might not want to leave any of it uneaten, and so we will look around the room, quickly count the number of mouths, and proceed to slice the cake accordingly. Four mouths? Four slices. Eight mouths? Eight slices. Eleven mouths? OK, that's a bit awkward when it comes to slicing. But you get the point.

What is most important from the perspective of this book is that at the most fundamental level, slicing and distributing the cake is a **sequential**[2] process which takes place one slice at a time. Each choice (i.e., each slice of cake) is connected in a somewhat hidden way to the choice that came before it, in the sense that the total amount that has been served up to a particular point in the sequence has implications for what remains to be served. In addition, each choice has implications for the decision that comes next. For example, if the initial slices are too large, the last guests to be served will get shortchanged. Or, if the initial slices are too small, some of the cake may be left at the end. Over time and with practical experience, we have developed an intuitive feel for slicing and serving cake. And because minor errors are often unimportant when it comes to serving cake, most of us tend to think about the entire problem before we even make the first cut. We might quickly count the number of guests and work out in advance how big the slices should be. In other words, we make the entire set or sequence of decisions **simultaneously**. As it turns out, that strategy turns out to be very useful, especially in simple problems, and can be setup in a rather uncomplicated mathematical way. In the case of allocating cake, we might even mark out the slices before beginning by putting small indentations in the icing. For a straightforward problem like this one, it is relatively easy to get our heads around the complete problem and solve it in a comprehensive way. Later in the book, we'll examine this comprehensive approach mathematically.

In recognition of the analogy at hand, the simplest dynamic optimization problems are often called **cake-eating problems**. One can apply the cake-eating logic to pumping oil from a well, extracting minerals from a mine, selling goods out of inventory, or spending down retirement savings. One begins with a resource of fixed size and simply tries to determine how best to allocate it over time.

---

progress by simply thinking of everyone's tastes as similar. Many economic problems focus on a single individual or a single firm, where interpersonal comparisons aren't necessary. But in many cases, our simplification is a dodge. See Baumol (1988).

[2] Throughout the book, attention will be drawn to key terms and concepts in the text by placing them, at first occurrence, in **bold** typeface.

Obviously, simple problems don't always contain realistic details, but much of the intuition developed in a simple problem carries over to more complicated settings. Nearly everything you need is there, and just about everything that is missing can be added.

Ingredients

The solution to a dynamic optimization problem consists of a **path** of values for the decision variable or variables in the problem. In the cake-eating problem, the path consists of a set of slices: slice 1, slice 2, slice 3, etc., up to and including the final slice. A **feasible path** is simply a list of the sequential decisions that are possible. Thinking a bit abstractly, if when we add up a set of imagined slices of cake, they don't exceed the size of the cake, then we say the path is feasible. If we imagine a set of slices that requires more cake than is available at the start, then that set of slices, i.e., that solution path, is infeasible. Note that a feasible path may not necessarily be the best path—that remains to be determined. For example, there may be multiple feasible travel routes from your home to your classroom or office, and some may be quicker or more desirable than others.

At this point, we haven't yet explicitly incorporated into our framework the notion of **optimality**. That will come soon enough. Nevertheless, we can still differentiate between a feasible path and an optimal path. A feasible path is possible. An **optimal path** is both feasible and better than any other feasible path. In the context of slicing and distributing cake, an optimal set of slices is one that we can't improve upon, in the sense that reallocating small bits of cake among guests, reducing the size of some slices and increasing the size of others, would disappoint some guests more than it would please others.

Sometimes in dynamic optimization problems, you will see decision variables referred to as **choice variables** or **control variables**. These terms are synonymous. They are simply the variables under the decision maker's control. They guide the underlying dynamic process. In most cases, a complete description of a dynamic optimization problem will also include a **planning horizon** that describes when the path starts and ends. The optimal path for such a problem then becomes a complete collection of values at each point in time or sequence stretching from the beginning to the end of the problem. The end of the planning horizon is typically fixed at a point in time, leading to what we call a **finite horizon** problem. In other words, we are told when the problem starts and when it ends. For example, a firm may need to report the value of inventory in the context of a tax year with a specified start and end date. In some cases, however, the horizon may be under the decision maker's control or, indeed, may extend forever. Although it seems difficult to imagine finding the solution to a problem that has no end, we will later see how a few mathematical tricks can help us solve **infinite horizon** problems. Of course, if we think about a party that lasts forever, we might need to concern ourselves with whether the cake is going to become stale or what value it might retain at the end of the problem. Fortunately, there are mathematical approaches to dealing with those situations, using a so-called **salvage value**. This, for example, might correspond to the value of an automobile at the end of its useful life, when it is "salvaged" by being crushed into a brick of metal and sold as scrap for recycling.

In describing dynamic optimization problems, one typically uses the term **stage** to keep track of the passage of time or to simply refer to the location or position of the system in time or sequence. The **initial stage** is the moment when the problem starts, and the **terminal stage** is the moment when the problem ends. The term **state** simply refers to the essential characteristics of the problem, in other words, the things we want to measure and keep track of. The state provides a way to describe the status of the system at any particular stage or point in time. We normally keep track of the state using one or more **state variables**. The state may characterize something about where things are starting, indicate what has happened in the past, and reveal something about what might be possible in the future. As an example, in the initial stage of the cake-eating problem, the state of the cake is "whole" or "uneaten." In the terminal stage of the cake-eating problem, the state of the cake is "all gone" or "eaten." Each stage of the problem has a corresponding state of the cake, i.e., a quantity of cake available. Between "whole" and "eaten," each intermediate state along the path results from a choice made in a prior stage of the problem. Stepping slightly outside the problem, you should get used to the idea that, even for feasible paths, there could be a large number of combinations of stages and states, some better in the context of the problem than others.

This all might seem like just a bunch of arbitrary words at the moment, but the definitions and underlying logic are important. At each stage, the choice made for the control variable has implications for the state variable and, as a consequence, for what is possible in subsequent stages of the problem. This basic fact is so important that it bears repeating:

> *Each state in a dynamic optimization problem results from the choices made in prior stages of the problem. Through their influence on the state variable, current choices have implications for what is possible in subsequent stages of the problem.*

This interconnectedness of states across stages is an essential feature of dynamic optimization. In the cake-eating problem, the amount of cake (i.e., the state) available at a particular point in the sequence depends on two things: how much cake was available in the prior stage (the prior state); and the size of the slice (the choice variable).

It is worth noting at this point that if a current decision has no meaningful implication for subsequent options and outcomes, then the optimization problem is not really dynamic, it is static. Although simply introducing time as a dimension isn't always enough to transform a static problem into a dynamic one, in economics, it is usually a good bet that a well-formed problem involving time will be a dynamic optimization problem.

Stages, States, and Friends

Returning to the cake-eating example, we can define the state as "the amount of unsliced cake available to serve" and the stage as "which guest is being served." We can think of the state as a variable that takes on values ranging from 1 (a complete cake available) to 0 (no cake remaining). Our choice variable, which is really a set of values, is the size of each slice. If I count myself among the guests, with eight people total to serve, then it is easy to see that there are eight stages (guests 1 to 8). We can use the stages to line up the eight decisions to be made (how much to serve each person) and the eight distinct observations on the state variable (the evolving size of the cake). Although we might think about this problem as unfolding over time, this isn't strictly necessary. In fact, in this problem, it doesn't matter who gets served first and who gets served last, as long as everyone gets served. We can line up the individuals alphabetically by name, by height, or by age. In all cases, the serving takes place sequentially.

Although it might seem tedious, let's lay out two possible versions of the cake-eating problem in tabular form. Example 1 is in Table 1.1.

In addition to providing an accounting of where the cake is going, the table entries relate what is possible (the current state) with what is selected (the choice) at each serving step (the stage). The final column also allows us to draw an accounting connection between what has just happened and what is about to happen. It demonstrates how the combination of the current state and choice influences the next state and the choices that will then be available. In each stage, we begin with whatever cake is available. We then allocate some cake to someone and subsequently consider how to allocate the remaining cake to everyone who has not yet been served. One subtle but important thing to note is that the state variable, which in this case is the amount of cake available, conveys everything we need to know about what has happened up to that point in the problem. There are many ways to end up in the middle of the problem with half a cake, but that doesn't really matter for purposes of deciding what is feasible along the remaining path. To convince yourself of this, consider example 2 in Table 1.2, which involves unequally-sized slices of cake.

Table 1.1 Cake-eating example 1: equal slices

Stage (guest)	State (cake available)	Choice (size of slice)	Subsequent state (cake remaining)
1. Richard	Complete cake	Slice = 1/8	7/8
2. Aiki	Cake = 7/8	Slice = 1/8	6/8
3. Carlos	Cake = 6/8	Slice = 1/8	5/8
4. Olivier	Cake = 5/8	Slice = 1/8	4/8
5. Pedro	Cake = 4/8	Slice = 1/8	3/8
6. Lisa	Cake = 3/8	Slice = 1/8	2/8
7. Diji	Cake = 2/8	Slice = 1/8	1/8
8. Jerry	Cake = 1/8	Slice = 1/8	0/8

Table 1.2 Cake eating example 2: unequal slices

Stage (guest)	State (cake available)	Choice (size of slice)	Subsequent state (cake remaining)
1. Richard	Complete cake	Slice = 2/8	6/8
2. Aiki	Cake = 6/8	Slice = 2/8	4/8
3. Carlos	Cake = 4/8	Slice = 0/8	4/8
4. Olivier	Cake = 4/8	Slice = 0/8	4/8
5. Pedro	Cake = 4/8	Slice = 1/8	3/8
6. Lisa	Cake = 3/8	Slice = 1/8	2/8
7. Diji	Cake = 2/8	Slice = 1/8	1/8
8. Jerry	Cake = 1/8	Slice = 1/8	0/8

In example 2, we serve Richard and Aiki large slices, we serve Carlos and Olivier nothing, and we serve everyone else the same portions as in example 1. The "paths" in example 1 and example 2 are both feasible, in the sense that they conform to the limitations laid out in the problem, beginning with a complete cake and ending with an empty plate. However, determining which path, if either, is optimal requires some additional analysis and requires us to carefully define what we mean by optimal.

Although it might seem obvious, it is worth repeating that different combinations of states and choices lead to different outcomes and different solution paths. In example 2, for instance, if we serve Pedro half of the cake in stage 5, which is feasible given the choices made in stages 1–4, then it becomes impossible to serve anything at all to Lisa or Diji (or, sadly, myself). Whether we think it is optimal to allocate 1/8 of the cake to each of the guests, or 1/2 of the cake to Pedro and 1/2 of the cake to someone else, depends a bit on how we define and measure optimality. This definition and measurement, in turn, should probably depend on what our guests like or dislike. If the cake is chocolate and Diji dislikes chocolate (which is unlikely, by the way), then it might be optimal and make sense to divide the cake in such a way that she is left with nothing.

Before going any further, it is useful to pause here and focus on a few other basic dynamic optimization concepts illustrated by this simple cake-eating example. Maybe the most important one to recognize is that, in this problem, the size of the cake is established and fixed at the start of the problem. No matter what we decide to do, we cannot grow the size of the cake. That's an important limitation. In fact, in many of the most interesting economic problems, it may be possible to grow the size of the cake over time. If our cake consists of capital or wealth, for example, foregone consumption at early stages may allow the cake to grow over time, providing greater opportunity for consumption at later stages. In a cake-eating problem, however, such opportunities are not available. It is only eat, eat, and eat.

Another key concept relates to what we call **initial conditions** and **terminal conditions**. Whenever we set out to formulate and solve a dynamic optimization problem, we need to be careful, explicit, and precise about what is happening at the start and end of the problem. In our cake-eating example, we start with a cake. That much is pretty clear. Before we serve anyone, we have a complete cake at our disposal. But what about the end of the problem? Thus far, we have somewhat implicitly assumed that we will serve all of the cake and therefore that at the end of the party, no cake will remain. Upon reflection, that seems like a fairly special case that might require either lots of guests or a very small cake.

In the language of dynamic optimization, we call these features **endpoint restrictions**. They define the start and finish of the problem and, in a strict sense, require us to be precise about defining the combination of the stage *and* the state at the beginning as well as the stage *and* the state at the end of the problem. Because dynamic optimization problems are concerned with choices along an optimal path, problem solving often hinges on the ways in which we define or constrain the endpoints for a problem. In theory, we can be flexible about both the initial (starting) and terminal (ending) stages of the problem. We can also be flexible about the initial or terminal states. In a sense, such flexibility creates four additional choice variables (i.e., (i) when to start, (ii) where to start, (iii) when to end, and (iv) where to end). In practice, however, most of the problems with which economists are concerned are fixed at the start by an initial stage and state. That is, we begin at a given time with a given initial value for the state variable. Even with the initial stage and the initial state fixed, however, we can still imagine a range of terminal stages and states. Our party could end at 5pm, 11pm, or the next day. That might have implications for how the cake gets eaten. Or we might want to reserve part of the cake for later, which is a restriction on the terminal state that may have implications for choices in the rest of the problem.[3]

More Interesting Economic Problems

Sequential decisions appear throughout economics. A classic example from macroeconomics is to find the optimal amount to save and reinvest out of production to maximize society's consumption over time. We will take up that problem in Chaps. 4 and 8, where we will see how savings decisions made today influence the rate of capital accumulation over time, thereby determining the set of feasible consumption and saving plans in the future. In fact, many of the earliest applications of dynamic optimization in economics were concerned with how an all-knowing and benevolent social planner might choose rates of savings and investment to maximize economic welfare over time. The early article "A Mathematical Theory of Savings" (Ramsey, 1928) is one of the first rigorous treatments of this topic. The question of what constitutes an optimal investment path and what influences an economy's movement along this path is a subject of perennial interest in macroeconomics and development economics.

---

[3] Beginning sometime in the nineteenth century in England, newlyweds started saving part of their wedding cake to consume on their first anniversary. Such a tradition continues to this day in many places, which has important implications for how much cake the wedding guests get to eat! If, in our example, we needed to save part of the cake for later, that binding terminal condition would require a change in allocation, and we would likely respond by dividing the remaining amount among the eight guests, with each receiving a slightly smaller slice than before.

A large number of microeconomic problems are dynamic as well. These include the problems of household savings and consumption, investment by a firm in research and development, and an individual's job search. Finding the optimal way to adjust a portfolio of investments over time is an example of a dynamic optimization problem that is complicated by uncertainty because the returns on investments, which themselves influence future asset allocation opportunities and hence future returns, are stochastic. We'll examine some problems like these in later chapters.

Dynamic optimization also applies to problems in environmental and natural resource economics. Consider the owner of a non-renewable mineral resource who must plan how much ore to extract from her mine over time. Ore extracted today can be sold, but if the resource is scarce and depletion leads to both higher costs of extraction and rising prices for the resource over time, then it might make sense to leave some of the ore in the ground for future sale. Another example comes from the analysis of groundwater extraction. If pumping costs rise as the water table drops but rainfall recharges the aquifer over time, then finding the optimal rate of extraction requires a delicate balancing between the benefits and costs of using water today and the alternative of saving some for tomorrow. Fisheries management is another area of application. Harvesting fish may generate economic returns, but harvesting all of the fish is sub-optimal because unharvested fish reproduce, creating future harvests. How might we balance harvesting now against harvesting later? This kind of question, which is ideally suited to the mathematics of dynamic optimization, goes to the heart of policy debates about how to achieve sustainable development.

Additional Considerations

To reiterate a point from above, unless decisions and outcomes are connected in some way across stages, a problem is not dynamic, even if time appears. That said, most problems in economics that involve time will, on close inspection, meet this test. Just to be clear on this point, although a series of static problems can be connected over time, such a sequence does not necessarily constitute a dynamic problem. In order for a problem to be truly dynamic, it must have the characteristic that it involves sequential decision-making in which current decisions have implications for subsequent opportunities, choices, and outcomes.

Similarly, although the inclusion of time in a problem should serve as a clue that the problem is dynamic, time is not a necessary component of dynamic problems. You also can think of dynamic optimization as a problem of multistage decision-making, in which steps must be carried out in a sequential or logical order, as if one were optimizing factory operations along an assembly line. Table 1.3 contains some examples of dynamic optimization problems in economics.

To emphasize the sequential nature of dynamic optimization problems, consider Fig. 1.1, which is a simple example of a four-stage network problem in which a sequential set of steps must be followed. In this case, imagine a traveler who needs to travel between location A and location J, with alternative routes that could be followed. The goal is to make the trip as inexpensively as possible. Although described

Table 1.3 Examples of dynamic economic problems

Economic scenario	State x	Control u	Benefit $F(x,u)$
Capital accumulation	Capital stock	Consumption	Utility of consumption
Mining	Ore deposit	Extraction rate	Profit from extracted ore
Selling an asset	Value of asset	Hold or sell (0/1)	Capital gain
Cutting timber	Tree biomass	Cutting (0/1)	Profit from cut timber
Harvesting fish	Animal biomass	Effort or catch	Value of catch
Raising livestock	Liveweight	Feed	Value of livestock
Irrigating a field	Water level	Water release	Crop yield or risk reduction
Controlling pests	Insect population	Pesticide application	Value of reduction in damage
Land degradation	Soil fertility	Crop or technology	Yield or profit
Groundwater usage	Pumping depth	Extraction rate	Value of water
Pollution reduction	Pollution level	Mitigation	Reduced harm
Product sales	Buyers	Advertising	Sales, profit, or market share
Inventory management	Stock on hand	Orders	Sales and storage costs
Final course grade	Current grade	Time studying	Utility of grade

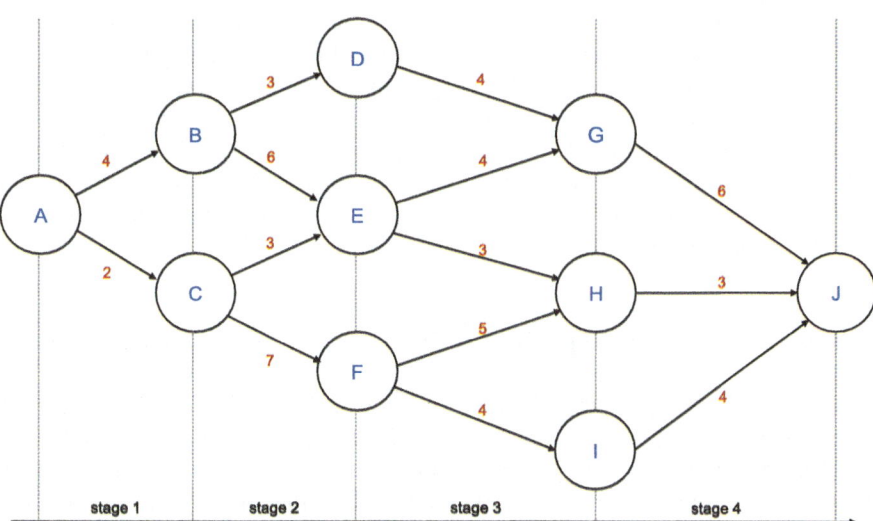

Fig. 1.1 A multi-stage travel problem from origin (A) to destination (J)

as a travel model, the setup is generic enough that it could also represent the transformation of raw materials into a finished product through various stages of production. In either case, one can think of each circle as representing a state. As illustrated, seven routes are available (without backtracking) starting from location A. This diagram is known as a **directed graph**. Each intersection in the graph is a **node**. Lines connecting nodes are **arcs,** and each arc is labeled with a numeric value associated with the cost of moving between adjacent nodes along that arc. Note that not all paths can be reached from each node. For example, choosing the path from A to B in stage 1 rules out the opportunity of moving from F to H in stage 3 (again, without backtracking).

This simple example illustrates the concept of **path dependence**, which is the idea that, along a dynamic path, final outcomes or feasible choices may depend on past decisions that cannot be undone. As a result, choices made in the initial stages of a problem (including sub-optimal choices) may preclude shifts to a different path (perhaps even a better path) at a later time. **Hysterisis** is a related concept and refers to the dependence of a state variable on the history of the system, perhaps due to path dependence, lags, or initial conditions.[4]

The goal of dynamic optimization solutions is to identify a path that is both feasible and optimal, where we define optimality using some maximization or minimization criterion. What path from A to J is optimal in Fig. 1.1? The answer depends on context and, in this case, the specific optimization criterion and the definitions of arc values. If the values along the arcs connecting nodes in Fig. 1.1 represent travel costs, and the goal is cost minimization, then some careful accounting will show that an optimal path is A–C–E–H–J, resulting in a total cost of reaching node J from node A of 11. On the other hand, if the circles represent value added and the goal is maximization of value added, then a different optimal path might emerge. Note that although there is a unique cost-minimizing path in this problem, the uniqueness of the optimal path is not always assured.

To find the optimal path from A to J, you might use brute force, simply listing every feasible path from A to J along with its associated value. With only seven feasible paths, this isn't hard to do. Another approach is to work in reverse, starting from J, using **backward recursion**. For lengthy or complicated problems, this second approach turns out to be more efficient. We will revisit this travel problem in Chap. 11 and solve it using backward recursion.

Example: A Thought Experiment Regarding Your Daily Commute

Let's close this chapter with a thought experiment. Consider one of your normal travel routines, say your daily commute. If you can, find or print a map of your city or town. Indicate on the map your normal travel route to school or work. Identify initial starting points, your final destination, and various decision points and the

---

[4] Path dependence is often invoked as the reason modern computer keyboards are laid out (suboptimally, from the perspective of the present) in the QWERTY pattern attributed to Sholes, Lidden & Soulé. According to Noyes (1983), keyboards on early mechanical typewriters may have been designed, in part, to slow down typists in order to prevent mechanical jams when tapping frequently-used keys in combination. We could type faster today with a different design, but making the switch seems almost impossible to contemplate, let alone implement (although some have tried: the Dvorak keyboard, which is probably the most famous attempt, was developed in 1936).

paths between them. What is your objective when traveling between home and school? Your normal method of transport? Does your favorite route depend on your method of transport? Try illustrating your route in abstract form using a diagram like Fig. 1.1, identifying the key nodes and arcs. Can you assign arc values along the paths connecting nodes? To what extent is your path constrained? Have you ever experienced a disruption of some kind that required a change in your plan or backtracking? If so, what did this do to your travel time? In what sense is traveling to school or work a dynamic optimization problem?

References

Baumol, W. J. (1988). *Superfairness*. The MIT Press.
Noyes, J. (1983). The QWERTY keyboard: A review. *International Journal of Man-Machine Studies*, *18*, 265–281.
Ramsey, F. (1928). A mathematical theory of savings. *Economic Journal*, *38*(152), 543–559.

Mr. Kuchenfresser

2

This chapter introduces Mr. Kuchenfresser. Through him, we will develop some examples to enhance our understanding of dynamic optimization.

Alternatives and Consequences

To get started, we'll need to simplify our simple cake-eating problem in one dimension but complicate it in another. As we do this, the concept of cake will become a bit more abstract, but the concept of dynamic optimality will become more concrete. The goal at this point is to develop intuition about how the mathematics works. This will allow us to expand our model in useful ways.

We also need to make a few assumptions. OK, actually, quite a few assumptions. First, let's suppose that there is only one cake eater—we'll call him Mr. Kuchenfresser. And only one cake to eat. Mr. Kuchenfresser has the opportunity to spread his cake eating over two days: today and tomorrow. We'll call today stage 1 ($t = 1$) and tomorrow stage 2 ($t = 2$). Second, let's assume more is better: Mr. Kuchenfresser can never get enough cake and, unlike Princess Lenore (in the James Thurber story "Many Moons"), he never falls ill from a surfeit of sweets. Third, let's assume that any cake eaten tomorrow is just as tasty as a cake eaten today. In other words, the quality of the cake itself doesn't change over time. Let's also assume that any cake left on the plate at the end of our planning horizon has no value. This could be because the cake immediately becomes stale and inedible at the end of day 2, or it could be because Mr. Kuchenfresser is going away on a long and well-deserved vacation to the Azores on the morning of day 3. We really don't need to be explicit about reasons. From our point of view, there is no stage 3 in the problem. Everything we care about takes place in stages 1 and 2.

We could get mathematical at this point, but before going any further, let's think about what a reasonable solution to this problem might look like.

© The Author(s), under exclusive license to Springer Nature Switzerland AG 2025   13
G. Shively, *A Beginner's Guide to Dynamic Optimization in Economics*, Classroom
Companion: Economics,
https://doi.org/10.1007/978-3-032-09374-5_2

One option would be for Mr. Kuchenfresser to eat all of his cake today. That leaves nothing for tomorrow, but, hey, why put off until tomorrow something you can eat today? We might call this the Garfield solution.[1]

A second option would be for Mr. Kuchenfresser to wait and eat all of his cake tomorrow. Spending today thinking about his cake might be torturous, but perhaps he could read a book on dynamic optimization to occupy his mind.

And in between these two options, there are an infinite number of alternatives from which to choose. How can we identify the best one?

In our search for a solution, we might want to think about Mr. Kuchenfresser's preferences. He's human after all, and our basic economic training tells us that humans largely obtain their utility from consuming. If we knew something about the shape of Mr. Kuchenfresser's utility function, would that information get us closer to a solution?

The answer, as in most economic circumstances, is: *it depends*.

To see why, let's construct a mathematical version of our problem. Let's call c_1 the amount of cake consumed in stage 1, c_2 the amount eaten in stage 2, and $u(c)$ the utility derived at a point in time from consuming c amount of cake. Mr. Kuchenfresser needs to solve the following problem:

$$Max\ u(c_1) + u(c_2)$$

$$\text{subject to}:\ c_1 + c_2 \leq C$$

where C represents the total amount of cake available. The inequality constraint indicates that the total amount eaten over the two days can't exceed the total amount available, but could be less. In fact, since we have assumed that Mr. Kuchenfresser can never get enough cake and that there is no reason to save cake beyond period 2, we can rewrite the inequality as an equality: the combination of today's consumption and tomorrow's consumption will always equal the amount of cake available, still leaving Mr. Kuchenfresser a bit peckish.[2] Equipped with this knowledge, we can form a Lagrangian function to solve the problem.[3] This can be written:

$$L = u(c_1) + u(c_2) + \lambda(C - c_1 - c_2).$$

Mr. Kuchenfresser's problem has two choice variables, c_1 and c_2. To find their optimal values, we derive the first-order necessary conditions for this maximization problem and set each of them equal to zero:

---

[1] For those unfamiliar with Garfield, he is a comic book cat created by Jim Davis. Garfield is noteworthy for his robust appetite and gluttonous behavior.

[2] I'm using the British meaning of the word *peckish* here (hungry) rather than the American meaning (irritable). Britain and America are "two countries, divided by a common language" according to the Irish playwright George Bernard Shaw.

[3] In case Lagrangian functions are unfamiliar, the appendix to this chapter provides a brief introduction. You can also consult Chiang (1974).

$$\frac{\partial L}{\partial c_1} = u'(c_1) - \lambda = 0$$

$$\frac{\partial L}{\partial c_2} = u'(c_2) - \lambda = 0.$$

Here, we use the shorthand notation $u'(\cdot)$ to represent the derivative of Mr. Kuchenfresser's utility with respect to cake, i.e., $\partial u / \partial c_t$, the marginal utility of consumption at time t.

Since the same λ appears in both expressions, we can equate them to obtain:

$$u'(c_1) = u'(c_2)$$

What does this tell us? In words, it says that to maximize his utility from cake consumption, Mr. Kuchenfresser should equate the marginal utility of day 1's cake consumption to the marginal utility of day 2's cake consumption.

Is this helpful?

Actually, it depends a bit on the underlying shape of Mr. Kuchenfresser's utility function. To see why, imagine that his utility function is linear. In that case, each additional unit of consumption brings the same amount of pleasure as the preceding unit, and the marginal utility of consumption is constant. In other words:

$$u(c) = c$$

and

$$u'(c) = 1.$$

Unfortunately, if this is an accurate representation of Mr. Kuchenfresser's preferences, then we are in trouble. (Actually, he is in trouble.) No unique solution will exist to this cake-eating problem. To see why, let the total size of the cake equal 10. If all of the cake is eaten on day 1 and none is eaten on day 2, then $c_1 = 10$ and $c_2 = 0$. This means $u(c_1) = 10$ and $u(c_2) = 0$. This allocation is feasible since $c_1 + c_2 \leq 10$. But, it provides a total utility of $u(c_1) + u(c_2) = 10$, which is the same as when none of the cake is eaten on day 1 ($c_1 = 0$) and all of it is eaten on day 2 ($c_2 = 10$). In fact, you can easily confirm that for any feasible allocation that falls between these extremes, total utility is the same. The non-uniqueness of the solution arises because utility is linear in consumption. Each unit, regardless of the period in which it is eaten, provides the same amount of pleasure as any other.

Curvature to the Rescue

Long ago, economists fell into this mathematical quicksand but also found a vine with which to pull themselves out: curvature. If you actually observe consumer behavior, you find that, when faced with an ever-greater amount of something, most consumers begin to tire of it. Utility, in other words, is not a linear function of consumption, but a non-linear function of it; and marginal utility, far from being constant, tends to diminish with greater levels of consumption. If we account for this concavity of the utility function, we might instead express Mr. Kuchenfresser's utility in a form that provides some of this necessary curvature, say:

$$u(c) = ln(c)$$

which implies:

$$u'(c) = \frac{1}{c}.$$

In this case, because *marginal* utility depends on the level of consumption, the optimality condition that requires us to equate marginal utilities is:

$$u'(c_1) = \frac{1}{c_1} = \frac{1}{c_2} = u'(c_2)$$

which requires that we equate the day 1 and day 2 allocations:

$$c_1 = c_2.$$

That's the only way to ensure optimality and make the first-order condition hold. Any other allocation makes marginal utility in one period greater than marginal utility in the other period, and that's not optimal. In other words, Mr. Kuchenfresser should consume half his cake today and half his cake tomorrow. You can quickly confirm that with these allocations, total utility is $ln(5) + ln(5) = 3.22$, which exceeds the utility of any alternative feasible allocation. For example, $1/10^{th}$ today and $9/10^{ths}$ tomorrow provides a total utility of $ln(1) + ln(9) = 2.20$, as does $9/10^{ths}$ today and $1/10^{th}$ tomorrow.

It is useful to observe that we arrived at this answer for two separate but related reasons: one economic and the other mathematical. The economic reason is diminishing marginal utility: the more cake Mr. Kuchenfresser consumes today, the less each extra bite means to him. On some abstract level, if we allow our imaginary man to project himself into the next day, he can envision that the extra bite will bring him greater pleasure when his belly is empty than when his belly is full. Today's man engages in a bit of cake trading with tomorrow's man, reallocating cake across time periods until each is satisfied and indifferent to further trades. Bond traders call this **arbitrage**, namely trading across markets or time periods to take advantage of different prices (in our case, marginal utilities) for the same asset (in this case, cake). Intertemporal arbitrage works in this example because we assume cake today and cake tomorrow are the same asset, and because we assume utility today and utility tomorrow are comparable.

The second force at work, the mathematical force, is curvature. The logic of intertemporal maximization leads us to reallocate consumption across time until we find points with similar slopes on the time-dated utility functions. As we shall see, this forms the basis for nearly all dynamic optimization solutions, whether they are two-period problems or infinite-horizon problems. The goal is always to find the best allocation of consumption or activity across time, i.e., the optimal time path. Additionally, there is a hidden take-away message here, namely that when economic problems lack curvature, as when Mr. Kuchenfresser's utility was assumed to be linear in consumption, problems are far less likely to have unique or easy to locate solutions. We will revisit the importance of this point in Chap. 9.

Equations of Motion

Elements of a problem indexed at different points in time, as with Mr. Kuchenfresser's today and tomorrow, are usually connected explicitly via an **equation of motion** (or what is sometimes called a **transition equation** or **intertemporal resource constraint**). An equation of motion describes the way in which a state variable changes or evolves over time, and is typically a general accounting expression that holds true for all stages of the problem. Equations of motion are central to dynamic optimization, and we will return to them in detail in Chap. 6. In the case of Mr. Kuchenfresser's two-period problem, the equation of motion is quite simple and can be constructed by rearranging the resource constraint. To form the equation, we begin by recognizing that the state variable for Mr. Kuchenfresser's problem is the size of the cake at the start of any stage. Let's call that x_t, where we already know that at the start of the problem, time $t = 1$, the size of the cake is C. So, $x_1 = C$. The equation of motion tells us how the size of the cake evolves between stages depending on Mr. Kuchenfresser's decision about how much cake to eat. Since, Mr. Kuchenfresser starts with a cake of size C, and eats the amount c_1 in the first period, the amount available at the start of period two is given by:

$$x_2 = C - c_1$$

which is the same thing as saying:

$$x_2 = x_1 - c_1.$$

Here, x is the state variable and c is the decision (or choice, or control) variable. In words, we say that the value of the state variable at the beginning of stage two is equal to the value of the state variable at the beginning of stage one, minus the amount consumed in stage one. Since we are working with discrete steps in time, it is helpful if we think of a stage in the problem as consisting of a series of three events: (i) we learn how much of the resource (our state) is brought forward from the previous stage and is available at the start of the period; (ii) we make a decision about how much of the resource will be consumed in the current period (our choice); and (iii) we pass the remaining amount of the resource into the next period. This remaining amount becomes the value of the state variable at the start of the next period.

If we modified the problem such that Mr. Kuchenfresser's flight was rescheduled and he ended up consuming his cake over three periods, the amount available in period three would be given by:

$$x_3 = x_2 - c_2.$$

At this point, the pattern should be clear. The general form for the equation of motion is:

$$x_{t+1} = x_t - c_t \tag{2.1}$$

where we denote the state variable by x and say that the value of the state variable in the next stage of the problem, x_{t+1}, equals the value of the state variable in the current period, x_t, minus current consumption c_t. In our cake-eating example, even though cake consumed and cake on the plate are one in the same, we explicitly acknowledge that the available stock of cake is the state variable, and the amount of cake consumed is the choice variable. Both are indexed by time, but the functional relationship between them does not depend on time. It is always "next x equals current x minus current c." That's important and characterizes what we call a **stationary** problem. In stationary problems, the values of variables will change over time, but the relationship that governs them will remain the same. In contrast, **non-stationary** problems are much more difficult to solve. In those, the underlying structure of the problem, say the characteristics of the equation of motion itself, evolves or changes over time. It is as if you began playing a game for which you know the rules, say *Monopoly*, only to learn in the middle of the game that someone has changed the rules. We won't encounter problems of that kind in this book, but if you continue in your study of dynamic optimization, you may come across them, and so it is useful to be aware and on the lookout.

Obviously, your imagination may want to run wild at this point as you conceive of problems with multiple state variables and multiple-choice variables. State variables for Mr. K's flight *en route* to the Azores might include latitude, longitude, altitude, and remaining fuel; control variables might include the amount of thrust applied to the left engine, the amount of thrust applied to the right engine, and the adjustment of various flaps, ailerons, and stabilizers. We would undoubtedly require multiple equations of motion to describe the plane's trajectory in response to chosen adjustments in the control variables.

Notation and Nomenclature

Dynamic optimization problems are typically concerned with decisions and outcomes that vary over time. **Continuous time** problems are those in which time varies continuously or instantaneously. **Discrete time** problems are those in which time moves across fixed intervals (e.g., days, weeks, quarters, or years). The exposition of continuous time and discrete time problems generally differs from a mathematical point of view: continuous time problems rely on **differential equations** to specify changes over time via the equation of motion, while discrete time problems use **difference equations** to track these changes.

When formally specifying a dynamic optimization problem, keep the notation clear. We tend to use t to index the stage of a dynamic problem that starts at some initial time usually denoted $t = 0$ or $t = 1$ and ends at a terminal time $t = T$. In a continuous-time problem, the time path for the variable x is $x(t)$; in a discrete time problem, the shorthand for the time path for the variable x is written $\{x_t\}$. Note that convention suggests the use of parentheses for continuous time problems and subscripts and brackets for discrete time problems (we will return to this in the appendix to Chap. 4). These are notational conventions only, however, and subject to some degree of alteration. Often, the time index can be omitted entirely to reduce clutter, especially where it is clearly implied by the context of the problem. Thus, in a dynamic problem, the symbols x and y might be understood as variables defined over time, without the addition of a time index. The superscript * is typically reserved to denote an optimal value or optimal path.

The proper specification of a dynamic optimization problem always includes a clear statement of the objective of the problem, specified in terms of either maximization or minimization. We require that the objective for maximization or minimization is a single value, and that it is therefore a function of the time paths of all choice variables. For Mr. Kuchenfresser, the value for maximization is his total utility, which consists of the simple addition of utility in each period. Total utility therefore becomes a function of his consumption path across two periods. In dynamic optimization, we typically refer to the object of maximization (or minimization) as a **value function**. In formal terms, the relationship between a feasible path and a value function is a mapping from the path to a real number. We refer to this type of mapping as a **functional** to distinguish it from a function (i.e., a mapping from real numbers to real numbers). The notation used for a functional is $V[y(t)]$. In words, we say that the value of V, which is a scalar, is a function of $y(t)$, which is a path consisting of multiple values. If Mr. Kuchenfresser chooses a different consumption path, he may end up with a different scalar value for overall utility. Note that a functional differs from a composite function. Here, the value function V is literally a function of $y(t)$: the entire time path of values. Substituting a new time path results in a new scalar value for V. The optimal scalar value for the problem, V^*, is the one generated by the optimal time path of values, y^*.

Appendix: Lagrangians

Joseph-Louis Lagrange, for whom the Lagrange method is named, was an Italian mathematician, astronomer, and physicist who lived from 1736 to 1813. This means he was a contemporary of fellow mathematician Leonhard Euler, the Scottish economist Adam Smith, and both Beethoven and George Washington. What an interesting time to be alive!

The basic idea behind Lagrange's method is to convert a constrained optimization problem (either maximization or minimization) into a form that allows us to apply the standard derivative test associated with an unconstrained problem. For example, you may be familiar with maximizing an algebraic expression and know that to find it's extreme value (either a maximum or minimum) you need to set its first derivative (the **gradient**) equal to zero and solve for x. This is the so-called first-order condition. Lagrange's method simply extends this basic approach to settings in which arbitrary constraints are placed on the optimization problem. In technical terms, the original problem is reformulated to incorporate both the gradient of the function and gradient of the constraint (or gradients, in the case of multiple constraints). This modified form of the problem, consisting of the combination of the gradient of the function plus the gradient of any constraints, is referred to as the **Lagrangian function** or, more simply, as the Lagrangian. This approach allows us to straightforwardly use the first-order conditions of the modified problem to find the constrained optimum.

In the most general one-variable case, if we are given an objection function, say:

$$y = f(x)$$

and want to maximize it subject to a constraint of the form:

$$g(x) = c$$

then the Lagrangian function can be written:

$$Z = f(x) + \lambda[c - g(x)]$$

where c is a constant, and we define λ as the **Lagrange multiplier**. Written in this way, Z can now be understood as a function of two variables, x and λ, and our problem is now to maximize Z using the necessary first-order conditions associated with it. These are:

$$Z_x = f_x - \lambda g_x = 0$$

and

$$Z_\lambda = c - g(x) = 0.$$

Solving these two equations ensures a solution to the constrained optimization problem. If the problem includes multiple constraints, these can be easily accommodated by defining a unique Lagrange multiplier for each.

To recap, the method can be summarized as follows: in order to find the maximum or minimum of the function subject to the equality constraint, first build the Lagrangian, which consists of the original (unconstrained) objective function plus any constraint equations accompanied by their Lagrange multipliers. Then set all partial derivatives equal to zero, including the partial derivatives with respect to the Lagrange multipliers. In situations where the function is defined over two or more variables, the solution strategy is the same. Here is a very simple and transparent example:

Maximize

$$y = x^2$$

subject to

$$x = 10.$$

The Lagrangian function can be written:

$$Z = x^2 + \lambda[10 - x].$$

The necessary first-order conditions are:

$$Z_x = 2x - \lambda = 0$$

and

$$Z_\lambda = 10 - x = 0.$$

By simple substitution:

$$x^* = 10$$
$$\lambda = 20$$

and

$$y^* = 100.$$

In economic contexts, λ is often referred to as the **shadow value** or **shadow price** of the constraint. It measures the amount by which the objective function changes when the constraint is slightly relaxed. To illustrate this, consider an example in which a firm's profit depends on the number of workers employed in production, with a constraint on the amount of available labor.

Maximize

$$\pi = 5000L - 25L^2$$

subject to

$$L = 10.$$

The Lagrangian function can be written:

$$Z = 5000L - 25L^2 + \lambda[L - 10].$$

The necessary first-order conditions are:

$$Z_L = 5000 - 50L - \lambda = 0$$

and

$$Z_\lambda = L - 10 = 0.$$

Once again, by simple substitution:

$$L^* = 10$$
$$\lambda = 4500$$

and

$$\pi^* = 47,500.$$

In this case, we can see that if the constraint on labor is relaxed, and the firm is able to add one unit of labor, its profit will increase by

$$\Delta\pi = 55,000 - 51,975 = 4475$$

which is the approximate value of λ.[4]

---

[4] It is approximate in this case because we are working with discrete values for L. If we allowed for infinitely small changes in L, the values for λ and $\Delta\pi$ would match exactly.

Reference

Chiang, A. C. (1974). *Fundamental methods of mathematical economics* (2nd ed.). McGraw-Hill.

Impatience

3

You may be eagerly thinking ahead, so why wait? Let's now consider the importance of **impatience**.

Discounting

Imagine your favorite rich uncle fatally crashes his vintage Porsche while taking a hairpin turn at excessive speed, leaving you in his will $100,000 and a pile of twisted metal. Rather than following in his footsteps by buying your own sports car, you decide to save the money for some future actual need. To keep the money relatively safe, and because you think you won't need the money for a while, you decide to purchase a Certificate of Deposit (CD) from a bank insured by the Federal Deposit Insurance Corporation (FDIC). Let's say you shop around and find a 5-year CD paying an annual interest rate of 5%. At the end of 5 years, your CD will be worth:

$$1000 \ (1 + 0.05)^5 = \$127,628.$$

Alternatively, you might ask the bank how much it would be worth to them today, if you promised to pay them $127,628 five years hence. The answer, of course, is that $127,628 in five years is today worth only:

$$\frac{\$127,628}{(1 + 0.05)^5} = \$100,000.$$

The future, larger, amount is worth comparatively less today. In other words, we **discount** the future value, putting it in terms that are comparable to the

© The Author(s), under exclusive license to Springer Nature Switzerland AG 2025
G. Shively, *A Beginner's Guide to Dynamic Optimization in Economics*, Classroom
Companion: Economics,
https://doi.org/10.1007/978-3-032-09374-5_3

present.[1] If I asked how much you would be willing to pay to receive $100,000 in five years, your answer would be:

$$\frac{\$100,000}{(1 + 0.05)^5} = \$78,353.$$

The reason is simply that you could put $78,353 in the bank today, and annual compounding would grow this amount to $100,000 in five years. We say the **current value** of $78,353 is $78,353, the **present value** of $100,000 received in five years is $78,353, and the **future value** of $78,353, if received in five years, is $100,000. Ignoring any risk or uncertainty about future outcomes, and assuming we don't actually need the money right now, we should be indifferent between these amounts. In terms of intertemporal financial comparisons, they are equivalent.

This approach to comparing economic values at different points in time—be they benefits, costs, or marginal utilities—is the standard approach in economics and, despite some ethical objections, has been with us for a while.

Just as the distance between parallel railroad tracks appears to narrow in the distance, to the economist, values seem to diminish mathematically when they accrue in the future. In fact, the power of discounting is such that even an extremely large sum pushed far enough in the future diminishes greatly, especially if the discount rate is large. If you remain unconvinced, consider $1,000,000 to be delivered at some future date. If received in 100 years, it will be worth just $85,000 today at a discount rate of 2.5% and just $7,500 at a discount rate of 5%. If our planning horizon extends to 200 years, which might not be unreasonable for a major capital investment with an extremely long life, say a bridge, our future million dollars would be worth a paltry $50 today.

Justification for discounting has often followed this logic, relying on the observation that, when it comes to consuming, individuals are impatient and slightly prefer current consumption over future consumption.[2] But what does impatience imply mathematically in the context of dynamic optimization?

---

[1] The origins of modern discounting, it turns out, can be found in the early days of maritime exploration when merchants safely on shore would extend credit to ship owners based on the value of goods onboard a ship. The amounts lent in anticipation of the delivery of goods were discounted based on the time a ship was expected to remain at sea and, presumably, the risk that the ship might sink or be pillaged by pirates before reaching safe harbor. See Tawney (1963).

[2] This observation of impatience is by no means recent. Commentaries on the myopia of individuals regarding the future can be found among the earliest writers one wishes to consult. Even Plato remarked upon it in his *Laws* (see Plato, 2016; Book XI). The English philosopher John Locke observed that individuals generally require a greater future reward to compensate for one passed-by in the present; and "remoteness" formed the fourth tenet of Jeremy Bentham's hedonistic calculus. An alternative view of why we might discount (see Domar, 1947) comes from the observation that capital is productive, in the sense that a hypothetical machine which produces machines (i.e., "capital") is worth more to us today than next year because of the additional machines it can produce while we are waiting for the future to arrive.

To see, let's go back to Mr. Kuchenfresser. Recall that c_1 is the amount of cake consumed in stage 1, c_2 is the amount eaten in stage 2, and $u(c)$ is the utility derived from consuming c amount of cake. But now, let us assume our stages are separated not by a day but instead by a year. Furthermore, let us assume Mr. Kuchenfresser is impatient: he slightly prefers to eat his cake now rather than later. In particular, let's suppose that he discounts future utility at a rate of 5%. We say that his **discount rate**, δ, is 5%, or 0.05. To simplify notation, it is customary to sometimes convert this discount rate into a **discount factor** via the formula $\beta = 1/(1 + \delta)$.

Finding an Optimal Path

Introducing the discount factor from above as a weight on stage 2's utility, Mr. Kuchenfresser now needs to solve the following problem:

$$Max \ u(c_1) + \beta u(c_2)$$

$$\text{subject to: } c_1 + c_2 = C$$

where C again represents the total amount of cake available. We again form a Lagrangian function for our problem. This can be written:

$$L = u(c_1) + \beta u(c_2) + \lambda(C - c_1 - c_2).$$

We again find the optimal values for c_1 and c_2 via the first-order necessary conditions, which imply $u'(c_1) = \beta u'(c_2)$. If we again express Mr. Kuchenfresser's utility in a concave form, say, $u(c) = ln(c)$ then the optimality condition requires us to equate marginal utilities, and therefore the year 1 and year 2 allocations, such that:

$$c_2 = \beta c_1.$$

As before, this allocation is the only way to satisfy the equality implied by the optimality condition. But now Mr. Kuchenfresser does not equate his allocations in each period. Instead, because he equates the marginal utility from today to the *discounted* marginal utility from tomorrow, he now expresses a slight preference for current consumption over future consumption. To be exact, if $\beta = 0.95$, then tomorrow's consumption will be valued at 95% of today's consumption.

It is instructive to see how discounting sets off a kind of chain reaction. Suppose Mr. Kuchenfresser plans to consume for three periods. His problem becomes:

$$Max \ u(c_1) + \beta u(c_2) + \beta^2 u(c_3)$$

$$\text{subject to} : c_1 + c_2 + c_3 = C$$

where β is squared for period 3 because the future value is discounted two periods back to the present: period 3 is discounted to period 2, and period 2 is discounted back to period 1. We again form a Lagrangian function for our problem. This can be written:

$$L = u(c_1) + \beta u(c_2) + \beta^2 u(c_3) + \lambda(C - c_1 - c_2 - c_3),$$

and the first-order necessary conditions can be written:

$$\frac{\partial L}{\partial c_1} = u'(c_1) - \lambda = 0$$

$$\frac{\partial L}{\partial c_2} = \beta u'(c_2) - \lambda = 0$$

$$\frac{\partial L}{\partial c_3} = \beta^2 u'(c_3) - \lambda = 0.$$

For $U(c) = ln(c)$ these imply:

$$c_2 = \beta c_1$$

$$c_3 = \beta c_2$$

and

$$c_3 = \beta^2 c_1.$$

Finding the exact optimal path of consumption is now straightforward, since we have three equations and three unknowns with which to work. Our resource constraint tells us that:

$$c_1 + c_2 + c_3 = C.$$

The easiest way to proceed is to make substitutions from above for c_2 and c_3. This gives us:

$$c_1 + \beta c_1 + \beta^2 c_1 = C$$

which implies:

$$c_1 = \frac{C}{1 + \beta + \beta^2}.$$

It is now only necessary to specify C, the initial endowment and substitute our value for β. If $C = 100$ (a complete cake) and $\beta = 0.95$ then:

$$c_1 = \frac{100}{1 + 0.95 + 0.9025}$$

which means the optimal path for consumption is:

$$c_1^* = 35.06$$

$$c_2^* = 33.30$$

$$c_3^* = 31.64.$$

Note that we've followed convention and used an asterisk to indicate that these consumption levels are optimal. Mr. Kuchenfresser consumes close to a third of his resources in each period, but slightly more than a third in the first period and slightly less than a third in the final period. These amounts are consistent with his rate of impatience.

Characterizing the Complete Solution

Two other calculations are required to characterize the complete solution to Mr. Kuchenfresser's dynamic optimization problem. First, the value of the objective function is:

$$V^*\{c_t\} = u(c_1^*) + \beta u(c_2^*) + \beta^2 u(c_3^*) = \ln(c_1^*) + \beta \ln(c_2^*) + \beta^2 \ln(c_3^*) = 10.$$

Second, it is easy to recover from the initial endowment and the consumption path the optimal time path for the state variable. This path consists of:

$$x_1^* = 100.00$$

$$x_2^* = 64.94$$

$$x_3^* = 31.64$$

$$x_4^* = 0$$

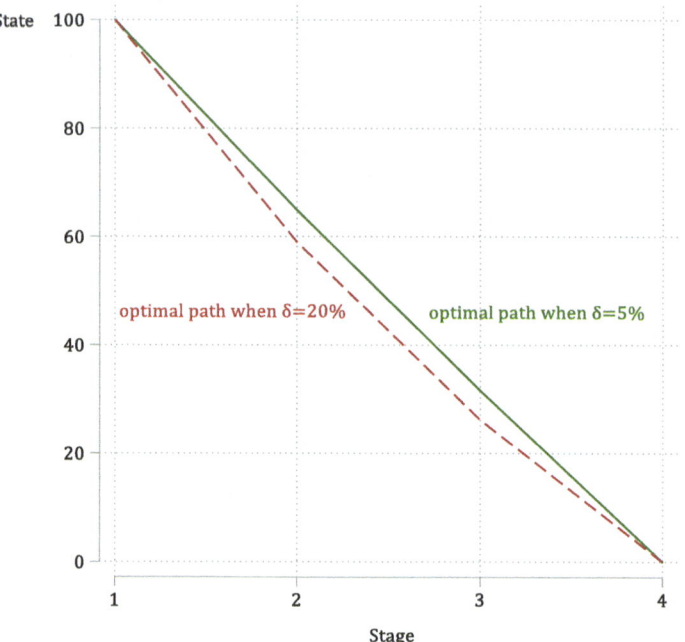

Fig. 3.1 Time path of Mr. Kuchenfresser's cake, the state variable

where we explicitly account for the fact that none of the resource remains at the end of the problem. That's fine, and optimal, since there is no value to be obtained from any resource in a hypothetical 4th stage of the problem.

It is easy enough to see that without these **time preferences** $\beta = 1$ and Mr. Kuchenfresser would consume exactly one-third of the endowment in each period. And the math confirms that a more impatient Mr. Kuchenfresser would shift even more future consumption into the present. For example, you can confirm that if $\delta = 20\%$, the optimal path for consumption is:

$$c_1^* = 40.98$$

$$c_2^* = 32.79$$

$$c_3^* = 26.23.$$

The resulting value of the objective function in this case is:

$$V^*\{c_t\} = \ln(40.98) + 0.80\ln(32.79) + (0.80)^2\ln(26.23) = 8.6.$$

This illustrates a basic feature of discounting, namely that higher discount rates imply greater weights on earlier consumption than later consumption. The implication is that, at the optimum, the optimal time path of consumption tends to favor earlier consumption over later consumption, and tends to bend the time path of the state variable towards earlier consumption. As Fig. 3.1 illustrates, with a larger discount rate we start and end at the same points, as indeed the resource constraint and endpoint restrictions require, but the drawdown of the state variable (cake) is more rapid at the start than when the discount rate is smaller.

References

Domar, E. D. (1947). Expansion and employment. *American Economic Review*, *37*(1), 34–55.
Plato. (2016). *Laws* (M. Schofield, Ed.; T. Griffith, Trans.). Cambridge University Press. (Cambridge texts in the history of political thought)
Tawney, R. H. (1963). *A discourse upon usury (1572)* (T. Wilson, Ed.). A. M. Kelley.

Mr. Kuchenfresser Meets Ms. Banker

4

We now push our example a bit further by adding savings alongside cake consumption, all the time generalizing our model to develop it more formally. In so doing, we are anticipating the tools of **optimal control theory**, which we will encounter in Chap. 8.

Adding Production and Savings

Recall from Chap. 2 Eq. (2.1), the general equation of motion for Mr. Kuchenfresser's problem. We wrote:

$$x_{t+1} = x_t - c_t$$

and said the value of the state variable in the next stage of the problem, x_{t+1}, equals the value of the state variable in the current period x_t, minus current consumption c_t. In our cake-eating example, even though the cake consumed and the cake on the plate appear to be the same, we explicitly acknowledged that the available stock of cake was the state variable and the amount of cake consumed was the choice variable. And we explicitly said the size of the cake could not grow.

Consider now the possibility of relaxing this last assumption and allowing uneaten cake to grow. If it isn't too weird to imagine, let's consider cake itself as a kind of fungible, all-purpose product that can be either eaten or left alone to grow into a slightly larger cake. Ms. Banker runs the process. There are several ways we can think about operationalizing this, but the easiest is just to imagine Mr. Kuchenfresser taking a slice from his cake to eat and saving whatever amount remains. Let's call the saved amount s_t. It is simple accounting to say:

© The Author(s), under exclusive license to Springer Nature Switzerland AG 2025
G. Shively, *A Beginner's Guide to Dynamic Optimization in Economics*, Classroom
Companion: Economics,
https://doi.org/10.1007/978-3-032-09374-5_4

$$s_t = x_t - c_t.$$

Mr. Kuchenfresser then drops off s_t with Ms. Banker, and in the next stage, she magically returns a slightly larger cake to Mr. K's doorstep. In other words:

$$x_{t+1} = (1 + r)s_t$$

where r is the rate of growth. Mr. K gets back his original saved portion of cake, s_t, plus the small increment rs_t. It is worth noting, as an aside, that this way of formulating the problem is completely consistent with the earlier, simpler treatment in which there was no way to grow the cake. In that case $r = 0$. It is also adaptable to situations in which $r < 0$ and our resource degrades over time. This depreciation of capital might be the case for, say, melting ice-cream or a product that declines in quality while in transit or sitting in a warehouse. You'll note that this setup (not coincidentally) looks a lot like a bank account, where interest is added to an asset left intact. How does this additional detail alter our approach?

A New Optimal Path

Let's go back and reconstruct Mr. Kuchenfresser's three-period problem. We'll continue to call c_1 the amount of cake consumed in stage 1, c_2 the amount eaten in stage 2, and c_3 the amount of cake consumed in stage 3. Remember that $u(c)$ is the utility derived at a point in time from consuming c amount of cake. We'll assume Mr. Kuchenfresser discounts his utility, that his utility function has some curvature, and that he wants to eat all of the cake available. He still wants to:

$$Max\ u(c_1) + \beta u(c_2) + \beta^2 u(c_2)$$

but now the total amount of cake available at the start of the problem isn't necessarily the full amount that will be available throughout the problem. Instead, any uneaten amount of c that is saved and carried over to the next stage of the problem is going to be larger than the amount saved. Let us use x_1 to represent the initial amount of cake and c_1 as the amount eaten in period 1. Then the amount saved (uneaten) is given by:

$$s_1 = x_1 - c_1$$

and the amount of cake available at the start of period 2 is:

$$x_2 = (1 + r)s_1.$$

As a result of this modification, Mr. Kuchenfresser still must choose how much to consume in each period and, just as before, anything not consumed in period 1 can be consumed in period 2, but now potential period 2 consumption receives an extra boost courtesy of saving foregone period 1 consumption.

At this point, your intuition should tell you that the presence of asset growth creates a modest incentive to save. You should also trust your intuition if it tells you that the larger is r, i.e., the rate of appreciation, the larger the incentive to save. As long as the utility function has some curvature, it will always make sense to spread consumption across periods. But now the optimal allocation across periods will depend on three competing forces: (i) the degree of curvature in the utility function, as represented by $u'(c)$; (ii) the individual's time preference, as represented by the discount factor β; and (iii) the return on savings, as represented by r.

Equipped with this complete set of knowledge, we can write out Mr. Kuchenfresser's problem as:

$$Max\ u(c_1) + \beta u(c_2) + \beta^2 u(c_3)$$

subject to the initial size of the cake, x_1, which is given as C, and the equations that link the size of the cake and the amount consumed to the amount saved in each period:

$$s_1 = C - c_1$$

$$s_2 = (1 + r)s_1 - c_2$$

$$s_3 = (1 + r)s_2 - c_3.$$

We're almost ready to derive the solution. But first, an additional observation is in order. Since we've specified in the setup that Mr. Kuchenfresser is only interested in his consumption for three periods, having any value for s_3, i.e., cake remaining for period 4, makes no sense. By that logic, s_3 should be equal to zero, ensuring that $c_3 = (1 + r)s_2$. In other words, the amount consumed in period 3 ought to be the full amount remaining.

This time, Mr. Kuchenfresser's problem has five choice variables: c_1, c_2 and c_3 plus s_1 and s_2. (Note that s_3 isn't included as a choice variable because we know it will be set equal to zero.) We can find the optimal values of the choice variables by deriving the first-order necessary conditions and solving the system simultaneously:

$$\frac{\partial L}{\partial c_1} = u'(c_1) = \lambda_1$$

$$\frac{\partial L}{\partial c_2} = \beta u'(c_2) = \lambda_2$$

$$\frac{\partial L}{\partial s_1} = \lambda_1 = (1+r)\lambda_2$$

$$\frac{\partial L}{\partial s_2} = \lambda_2 = (1+r)\lambda_3.$$

A bit of algebra applied to these yields:

$$\frac{u'(c_1)}{(1+r)} = \beta u'(c_2)$$

and

$$\frac{u'(c_2)}{(1+r)} = \beta u'(c_3).$$

In this case, to maximize his utility from cake consumption, Mr. Kuchenfresser should do something similar to the "no growth" case and equate the marginal utility of period 1's cake consumption to the discounted marginal utility of period 2's cake consumption, where that marginal utility is now weighted by how much growth results between periods from saving. If $r = 0$, i.e., if there is no growth in savings, then the problem reduces to what we saw in Chap. 3. With any positive r, however, there is an incentive to forgo a bit of consumption in this period in order to boost consumption in the next period.

A Numerical Example

To see this clearly, let's try a numerical example using $U(c) = ln(c)$, $r = 0.10$, $\beta = 0.95$ and $C = 100$. To begin, we can write out Mr. Kuchenfresser's problem as:

$$Max \ ln(c_1) + \beta ln(c_2) + \beta^2 ln(c_3)$$

subject to the initial size of the cake, x_1, which is given as $C = 100$ and the given savings rate of $r = 0.10$. As before, we use the equations of motion that link the size of the cake and the amount consumed to the amount saved in each period:

$$s_1 = 100 - c_1$$

$$s_2 = 1.1s_1 - c_2$$

$$1.1s_2 = c_3$$

where in the last case we've incorporated the fact that having any value for s_3 remaining for period 4 isn't optimal.

We can once again find the optimal values of the choice variables by deriving the first-order necessary conditions and solving the system simultaneously:

$$\frac{\partial L}{\partial c_1} = \frac{1}{c_1} = \lambda_1$$

$$\frac{\partial L}{\partial c_2} = \frac{0.95}{c_2} = \lambda_2$$

$$\frac{\partial L}{\partial c_3} = \frac{(0.95)^2}{c_3} = \lambda_3$$

$$\frac{\partial L}{\partial s_1} = \lambda_1 = 1.1\lambda_2$$

$$\frac{\partial L}{\partial s_2} = \lambda_2 = 1.1\lambda_3.$$

These imply:

$$\frac{1}{1.1c_1} = \frac{0.95}{c_2}$$

$$\frac{1}{1.1c_2} = \frac{0.95}{c_3}$$

$$c_3 = 100 - c_2 - c_1.$$

With three equations and three unknowns, we can solve by substitution to obtain:

$$c_1 \approx 31.89$$

$$c_2 \approx 33.33$$

$$c_3 \approx 34.78.$$

Intuitively, we can see that consumption is not the same in all three periods. In particular, the fact that any savings (i.e., forgone consumption) grow, leading to

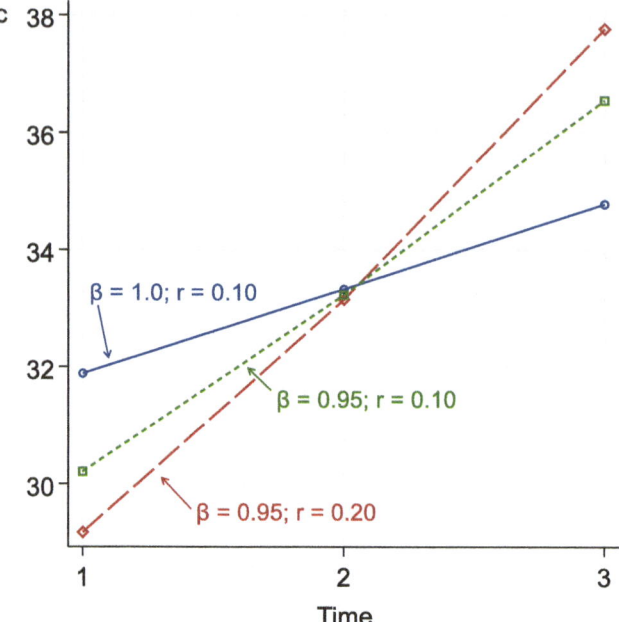

Fig. 4.1   Three consumption time paths with savings

potentially higher future consumption, creates a slight incentive to save in the first
period in order to consume a bit more in the third period. Of course,
Mr. Kuchenfresser continues to discount future consumption, so the shift isn't
very pronounced, but as long as the rate of growth in savings exceeds the rate of
discount, as it does here, the shift will be toward saving early and consuming
slightly more at a later date. Total consumption is the same, but the shift maximizes
the total discounted utility of consumption, as you can confirm.

To further convince yourself of the tug-of-war taking place between the savings
rate and the discount rate, let's solve the problem again, this time without discounting (i.e., with $\beta = 1.0$), to obtain:

$$c_1 \approx 30.21$$

$$c_2 \approx 33.23$$

$$c_3 \approx 36.55.$$

In this case, as expected, Mr. Kuchenfresser forgoes even more early consumption
in favor of later consumption.

And if we restore discounting ($\beta = 0.95$) but boost the return on savings (to
$r = 0.20$), the shift is even greater:

$$c_1 \approx 29.18$$

$$c_2 \approx 33.15$$

$$c_3 \approx 37.77.$$

Figure 4.1 illustrates these three consumption time paths.[1] What may or may not be apparent is that this solution method, which relies on building the chain of connections in consumption across time and then solving all of them simultaneously, works fairly well in a small problem like this. In the case of a larger problem, say with additional state variables or many time periods, the math becomes much more cumbersome.

Appendix: Standard Notation

When one formally specifies a dynamic optimization problem, the following notation is useful and fairly standard. We use t to index time, i.e., the stage of a dynamic problem which starts at initial time 0 and ends at terminal time T. As noted earlier, convention suggests the use of parentheses for continuous time problems and brackets and subscripts for discrete time problems. See Table 4.1:

These are conventions only and subject to some personal preference as long as one maintains consistency. Often, the time index will be omitted to reduce clutter, especially where it is clearly implied by the context of the problem. Thus, in a problem which is clearly dynamic, the symbols x and y might represent a vector of values for a variable which is defined over time, without the addition of a time subscript or any parentheses or brackets. The superscript * or the addition of a hat (as in $\hat{u}$) should always be reserved to denote an optimal (or solution) value (or path). Quite frequently, x is used to represent a state variable and u is used to represent a choice (control) variable, although the context of the problem should always be your guide to proper interpretation of letters, whether they are Latin or Greek.

Table 4.1 Typical notation in discrete and continuous time problems

Problem component	Time	
	Discrete	Continuous
Initial (start) time	t_0	$t(0)$
Time path for the variable x	$\{x\}$	$x(t)$
Initial value for the variable x	x_0	$x(0)$
Terminal value for the variable x	x_T	$x(T)$

---

[1] This is probably a good time to review notation. See the appendix to this chapter.

Euler, Euler, Master of Us All

5

The main difference between static optimization and dynamic optimization is that instead of dealing with a differential dx, which measures changes in the value of $y = f(x)$, we must instead deal with a shift or variation of the curve $y(t)$, which consists of many contributing pieces. For example, in solving Mr. Kuchenfresser's various consumption problems in previous chapters, meeting the goal of maximizing his aggregate utility of consumption (a single value) required choosing an entire set of consumption values along a path. The optimization took place with respect to the entire set of consumption choices. We couldn't choose one in isolation without understanding its impact on those that followed. In a static, single-variable calculus problem, the maximum or minimum of a function occurs at a critical point where the derivative is equal to zero. As mentioned at the end of Chap. 2, in a dynamic problem, we are looking for the maximum or minimum of a functional. A functional uses an entire function as its input and returns a single number.

Calculus of Variations

Technically, we say that a variation in the curve $y(t)$ arising from a shift in that curve produces a change in the composite value of the function $V[y(t)]$. Although this might sound a bit obtuse, we already encountered this concept at the end of Chap. 3. Recall that we derived two optimal paths for Mr. Kuchenfresser's consumption over time, one for a discount rate of 5% and another for a discount rate of 20%. Figure 3.1 was used to illustrate these paths. In the first case, $V[y(t)] = 10.0$, based on the consumption path {35.06, 33.30, 31.64}. In the second case, $V[y(t)] = 8.6$, based on the consumption path {40.98, 32.79, 26.23}. The value of the functional in each case depended on Mr. Kuchenfresser's entire consumption path and his discount rate. You might think of potential variations in these paths as

G. Shively, *A Beginner's Guide to Dynamic Optimization in Economics*, Classroom Companion: Economics,
https://doi.org/10.1007/978-3-032-09374-5_5

alternatives that give rise to higher or lower overall utility. Of course, we cannot directly compare the two results in Fig. 3.1 because we used different discount rates when generating them, and therefore were essentially solving slightly different optimization problems. In fact, both paths are optimal for Mr. Kuchenfresser, given the discount rate applied in each situation.[1] Still, you can see the intuition behind functionals and variations in curves in what we were doing. In both cases, we could adjust Mr. Kuchenfresser's consumption path while still meeting his endpoint restrictions, to see if we might improve his value function. We could have even done this by trial and error, say by starting in each case from the path {33.33, 33.33, 33.33} or the path {100, 0, 0} and adjusting the path in small steps. That kind of brute force approach is slow and tedious. Fortunately, calculus allowed us to home in on the optimal path more directly.

This focus on variations of a function is the origin of the label **calculus of variations**. The calculus of variations is the earliest form of dynamic optimization. It originated in the 1700s with Sir Isaac Newton and the Bernoulli brothers, and was perfected by the Swiss mathematician Leonard Euler, who is regarded as the most prolific mathematician of all time.[2]

In the remainder of this chapter, we devote our attention to the calculus of variations. To do so, we will need to move away from the example of Mr. Kuchenfresser, or at least begin to think about eating as a continuous process of nibbling crumbs and time as divisible into an infinite number of small increments. This is necessary because the techniques used in the calculus of variations are based on classical methods of calculus and rely on using first and second derivatives to find solutions. For this reason, the technique is restricted to problems for which all functions involved in the problem are continuous and continuously differentiable. These assumptions turn out to be very restrictive. In response, two other branches of dynamic optimization, **optimal control theory** and **dynamic programming**, have arisen to relax these restrictions. In this sense, the calculus of variations deals with special cases of problems that can be examined using the other methods. We will look at those other approaches in subsequent chapters.

The Euler Equation

The basic form of the calculus of variations problem is to:

---

[1] You can confirm with a bit of arithmetic that substituting each path into the alternative problem yields a solution with a slightly lower value for Mr. Kuchenfresser's objective function: using the second path for the first problem results in a value of 9.98 (vs. 10.00); and using the first path for the second problem results in a value of 8.57 (vs. 8.60).

[2] Reportedly, the French mathematician Laplace was fond of encouraging young mathematicians to "read Euler, read Euler, he is our master in everything." You can find this and other tidbits about Euler on his Wikipedia page. It makes for very inspiring (or humbling) reading. Note that one correctly pronounces *Euler* like *oiler*.

$$Max_{\{x\}} V[x(t)] = \int_{t_0}^{T} F(x(t), x'(t), t)dt \tag{5.1}$$

subject to the endpoint restrictions:

$$x(t_0) = x_0 \tag{5.2}$$

$$x(T) = x_T. \tag{5.3}$$

Let's break this down.

Equation (5.1) indicates that we are maximizing, and that in order to do so, we will be choosing a set of values for x. The endpoint restrictions, (Eqs. 5.2 and 5.3), simply define the start of the problem (time t_0) and the end of the problem (time T). They state that the path we choose for x must connect these endpoints. In Mr. Kuchenfresser's case, those would correspond to a full cake at the start ($x_0 = 100$) and no cake at the end ($x_T = 0$).

Turning back to (Eq. 5.1), $V[x(t)]$ is our functional: V is what we are trying to maximize. It takes on a single scalar value, which depends on the path $x(t)$. It is defined here in the most general way, that is, as the accumulation of values across a time path, where at each point in time we accumulate values according to the function, $F(\bullet)$, the integrand in (Eq. 5.1). At the moment, this function is formless, but we know that it's value may depend on the path, $x(t)$, on minor adjustments or variations to that path $x'(t)$, and possibly on t itself since, for example, a longer path could result in more time to accumulate value. Importantly, the objective must be optimized with respect to all three terms. If any of the three can be adjusted to give rise to a larger aggregate value for V, then we've done something which is sub-optimal.[3] Although we are currently working with a single path variable x, generalizing the Euler equation to the case of more than one path variable is possible.

To review, in the most straightforward calculus of variations problem, both the initial point (t_0, x_0) and the terminal point (T, x_T) are fixed. We define an **admissible function** (much like a feasible path) as any function on the interval $[t_0, T]$ that satisfies the endpoint restrictions $x(t_0) = x_0$ and $x(T) = x_T$. As with Mr. Kuchenfresser's problem, there may be a large number of admissible functions in the solution space, each of which describes a possible path: eat everything now, eat everything later, and everything in-between. The goal is to find an admissible function $x(t)$ that generates the largest value for $V[x(t)]$. We assume F is continuous in all of its arguments; and because the calculus of variations relies on classical methods of calculus, we restrict our attention to relative optima, i.e.,

---

[3] Parallel with the way we treat static problems, a dynamic minimization problem differs from the dynamic maximization problem, only with respect to the second-order condition. In the calculus of variations, this is referred to as the **Legendre condition**, named after the French mathematician Adrien-Marie Legendre, who outlined it in 1786.

those optima in the neighborhood of admissible paths. Note that in a calculus of variations problem, our equation of motion (Eq. 2.1 from Chap. 2), i.e., the constraint on adjustments to the state variable over time, is expressed by $x'(t)$, and is imbedded directly in the objective function. The importance of this will become apparent. Just think of $x'(t)$ as minor adjustments in x.

The first-order condition for optimality in the calculus of variations is the **Euler equation**, which is written as:

$$F_x = \frac{d}{dt} F_{x'}. \tag{5.4}$$

The solution path to the Euler equation is $x^*(t)$, which is the analog to the solution point x^* in a static optimization problem. In a static optimization problem, we would find x^* by setting $f'(x) = 0$ and solving for x. Our work in a dynamic problem is made slightly more difficult because a change in x at any point along the path has implications for outcomes along the portion of the path that remains.

Interestingly, the static problem with which we are all familiar is imbedded in the dynamic problem as a special case: if $dF_{x'}/dt = 0$, then the path is unaffected by time (i.e., its derivative with respect to t vanishes) and the Euler equation collapses to $F_x = 0$, which aligns with the standard static condition for optimality. In general, however, the objective in the dynamic problem must be maximized with respect to each term in the function, i.e., $x(t), x'(t)$, and t. Furthermore, the condition $F_x = \frac{d}{dt} F_{x'}$ must hold at every point along the path. Proving the optimality of the Euler equation isn't particularly difficult, but it is somewhat tedious. Interested readers can consult any number of sources for a proof, including (Benveniste & Scheinkman 1982, Kamien & Schwartz 1991, or Chiang 2000).

Some Intuition

This is a new way of thinking, and so, at this point, you may want to re-read everything written above about the Euler equation, sleep on it, and let it sink in as your brain builds new synapses. Once you have, eat a good breakfast, brew some strong coffee, and sharpen a box of Dixon Ticonderoga #2s (the world's finest wooden pencils). It's time to work on our intuition.

The left-hand side of Eq. (5.4), F_x, is the instantaneous change in the function of interest that arises from a change in x. The right-hand term, $dF_{x'}/dt$, is the rate of change (over time) of the function. The Euler equation states that, at the optimum, any instantaneous change in the value of the function must be balanced by the rate of change in the value of the function over time. We can interpret this in terms of Mr. Kuchenfresser's dilemma. Eating an extra bite of cake today provides an instant increment in utility. But once that bite has been taken, all subsequent bites will have to be reduced by a small amount to compensate (and thereby meet the overall resource constraint). The Euler equation tells us that, at all points along the optimal path, immediate changes in the objective function (measured by

F_x) must offset the impact of necessary adjustments (measured by $dF_{x'}/dt$). In the case of Mr. Kuchenfresser, this means instant gratification must make up for subsequent regrets. If these two things are not in balance, then it is possible to reallocate consumption in some way across the path to generate higher overall utility. If reducing consumption by a tiny bit at an early time period leads to greater overall satisfaction, because spreading out that extra consumption over the time remaining in the problem is a bit more valuable in terms of overall utility, then consumption should be delayed. Only when we become indifferent regarding adjustments can we say that we've found the optimal consumption path. This was the insight we gained in Chap. 4 when we considered interest on savings alongside discounting of consumption.

If the idea of instant gratification compensating for subsequent regrets is still a bit too abstract for you, consider a different analogy. Imagine for a moment that it's a hot summer day and Mr. Kuchenfresser has decided to head to the beach to cool off. Unfortunately, he's a poor swimmer and immediately begins flailing.[4] Fortunately, an attentive lifeguard on the beach spots Mr. Kuchenfresser drowning and immediately rushes to his aid. The lifeguard, of course, wants to reach Mr. Kuchenfresser in the shortest possible time and must therefore find the optimal path to minimize the total time needed to reach our desperate friend. This path will include a mixture of running and swimming, and we can assume our guard generally runs faster on sand than she can swim in the water. The optimal path for the lifeguard, therefore, is to enter the water exactly where an extra step along the beach provides no saving in total travel time. This is exactly what the Euler equation requires: the derivative of the rate of change in the functional must be zero with respect to variations in the function being optimized. Not too much time on the sand, but not too little.

Now, let's get technical about it. Note that F in Eq. (5.1) contains three arguments (x, x', t). Hence, the total derivative $dF_{x'}(x, x', t)/dt$ also will contain three terms. The total rate of change of the value of the function $F_{x'}$ depends on t directly and also indirectly through x and x'. By the chain rule:

$$\frac{dF_{x'}}{dt} = \frac{\partial F_{x'}}{\partial t} + \frac{\partial F_{x'}}{\partial x}\frac{dx}{dt} + \frac{\partial F_{x'}}{\partial x'}\frac{dx'}{dt}. \tag{5.5a}$$

We can rewrite the right-hand side of this more compactly as:

$$F_{x't} + F_{xx'}x'(t) + F_{x'x'}x''(t). \tag{5.5b}$$

Substituting this back into (Eq. 5.4), the original Euler equation, leaves us with:

---

[4] Note that this is *not* due to his over-indulgence in cake and failure to wait an hour after eating before going for a swim, as his grandmother instructed. According to the American Red Cross, the idea that swimming soon after eating increases the risk of drowning lacks empirical support.

$$0 = F_{x't} + F_{xx'}x'(t) + F_{x'x'}x''(t) - F_x. \tag{5.6}$$

Equation (5.6), the so-called **expanded form** of the Euler equation, is a **second-order differential equation**. Those who have encountered differential equations will know that the general solution to a second-order differential equation includes two-arbitrary constants. These constants are typically determined in a calculus of variations problem using the endpoints x_0 and x_T. In the lifeguard example, x_0 is the exact location of the lifeguard's chair, and x_T is Mr. Kuchenfresser's position in the water.

To preview coming attractions, it turns out that dealing with second-order differential equations can be a challenge. In some problems, therefore, it is easier to introduce a new variable, $\lambda(t) = F_{x'}(x, x', t)$. Once that's done, one can define the **Hamiltonian**:

$$H(x, \lambda, t) = F(x, \lambda, t) + \lambda x'. \tag{5.7}$$

This modification allows us to pull the equation of motion out of the integrand and re-cast the original problem in terms of two **first-order differential equations**. The Hamiltonian forms the core of solution methods in optimal control theory. This simplification is just one of the advantages of optimal control theory over the calculus of variations. However, much of the early literature on economic dynamics was developed using the calculus of variations, and so as an economist, it is useful to gain a basic familiarity with it in order to more easily read and understand some important, classic papers. For the moment, we postpone a discussion of Hamiltonians and turn our attention to some numerical examples of calculus of variations problems.

A Numerical Example

We can use a simple example, devoid of economic meaning, to outline the mechanics of solving a calculus of variations problem. Once we master this, we will turn our attention to a problem with economic content.

We want to find the optimum of the functional:

$$V[x(t)] = \int_0^4 \left(24tx + (x')^2\right) dt$$

subject to the initial condition $x(0) = 0$ and the terminal condition $x(4) = 64$. Remember, the value of $V[x(t)]$ will be a scalar: it is the accumulation of values from $t = 0$ to $t = 4$, just as Mr. Kuchenfresser's utility was a single composite value based on consumption over three time periods. Here, to find the optimal solution, we simply locate the parts necessary to construct the expanded Euler equation.

The objective function is the integrand:

$$F = 24tx + (x')^2$$

In a fairly mechanical way, this provides the necessary ingredients:

$$F_x = 24t$$

$$F_{x'} = 2x'$$

$$F_{x'x'} = 2$$

$$F_{x'x} = 0$$

$$F_{tx'} = 0.$$

We can now assemble these pieces to obtain the Euler equation. This is:

$$2x'' - 24t = 0$$

or, after dividing through by 2:

$$x'' - 12t = 0.$$

Our solution function must be an expression in x. Therefore, we need to integrate twice. The first integration gives:

$$x' - 6t^2 + z_1 = 0$$

where the first constant of integration is z_1. For those unfamiliar with solving differential equations, this is simply a placeholder at the moment. It will be necessary to solve for z_1 in a subsequent step. Integrating a second time gives the general solution:

$$x - 2t^3 + z_1 t + z_2 = 0$$

where z_2 is our second constant of integration. To find the constants of integration, we use the initial and terminal conditions provided in the statement of the problem. Using $x(0) = 0$ we set $x = 0$ at $t = 0$:

$$0 - 2(0)^3 + z_1(0) + z_2 = 0.$$

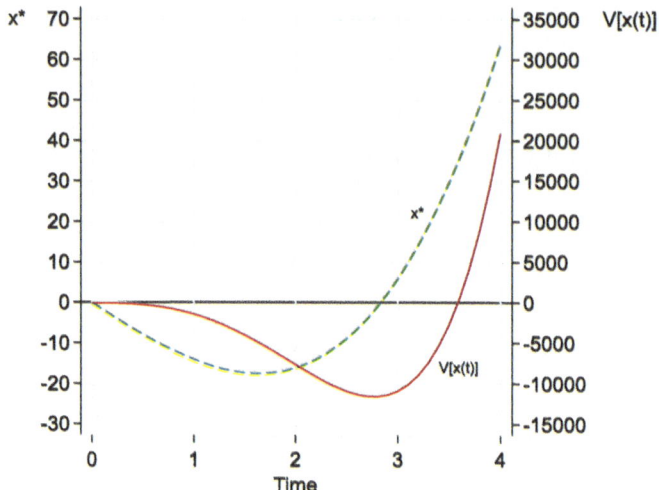

Fig. 5.1 The time path for $x^* = 2t^3 - 16t$

This gives us $z_2 = 0$. Now use the other endpoint to set $x(4) = 64$:

$$64 - 2(4)^3 + z_1(4) + 0 = 0$$

to obtain $z_1 = 16$.

The specific solution to the problem is therefore the cubic:

$$x^* = 2t^3 - 16t.$$

Figure 5.1 illustrates the time path x^* takes from $t = 0$ to 4, and the accumulating value of the value function along the path.

A Canonical Economic Problem

The following is a canonical economic problem that forms the basis for much theoretical and applied economic work.[5] Rather than simply wave goodbye to Mr. Kuchenfresser at this point, however, we'll endow him with a bit more economic agency. He is no longer simply "eating cake."

---

[5] In mathematics and economics, the term "canonical" means "according to standard," i.e., a classic example or form. The origins of the word are found in the Latin word for "rule," which itself comes from the ancient Greek word κανών (*kanon*), which was a straight measuring rod used by ancient builders to ensure consistency in their work.

Instead, Mr. Kuchenfresser seeks a consumption rule to maximize the discounted stream of his utility over a horizon of T years. He earns an exogenous wage w and an interest rate i on any savings that he does not consume. To acknowledge the productivity of these savings for future consumption, we label Mr. K's savings as capital, k.[6] We assume his utility is time separable and defined only over consumption of a composite good c. Utility $U(c)$ is strictly concave (i.e., $U' > 0$ and $U'' < 0$). Mr. K discounts his future utility at the rate δ.

Mr. Kuchenfresser's problem is now to maximize the discounted sum of utility from consumption:

$$Max \int_0^T e^{-\delta t} U(c(t)) dt$$

subject to changes in his invested capital, if any. We write the equation of motion for the capital stock as:

$$ik + w = c + k'.$$

This simply says that income (i.e., wages plus interest on any savings) must equal consumption plus net investment. In other words, any increment to capital (positive or negative) will equal interest income plus wages minus consumption. Most of the intuition for the solution comes directly from the equation of motion. Income leads to consumption, but if consumption eats into savings, then future consumption will suffer. This may sound very familiar based on our experience with Mr. Kuchenfresser at the end of Chap. 4. Just substitute r for i, s for k, and drop the w, and it begins to look much the same, except now we are dealing with a continuous time problem, and hence differential equations, rather than a discrete time problem governed by difference equations.

This is a calculus of variations problem in $k(t)$. We have the following expressions:

$$F_k = e^{-\delta t} U'(c) i$$

and

$$F_{k'} = -e^{-\delta t} U'(c).$$

---

[6] Economists use k to represent capital to avoid confusion with consumption, which is normally denoted c. The German word for capital is *Kapital*, which also helps explain the choice, since one can assume Mr. Kuchenfresser might speak German.

The Euler equation is therefore:

$$\frac{d}{dt}\left(-e^{-\delta t}U'(c)\right) = e^{-\delta t}U'(c)i.$$

If we differentiate and collect terms, we get the solution directly:

$$\frac{-U''c'}{U'} = i - \delta \tag{5.8}$$

where c' is shorthand for dc/dt. Note the similarity to the problem we solved in Chap. 4. We can interpret the solution as follows. At the optimum, the proportionate rate of change in marginal utility must equal the difference between the rate of interest on investment and the rate of individual impatience. Think of this as a "no arbitrage" condition for the individual. We know $-U''/U' > 0$ (because we've assumed the utility function is concave), so we can reason as follows:

If $i > \delta$, Mr. Kuchenfresser can be said to be patient. In other words, he discounts at a rate that is lower than what can be earned on savings in the bank. As a result, he prefers to save a bit. As a result, $dc/dt > 0$. In this case, Mr. K's consumption tends to rise over time because he tends to save early in his planning horizon and consume later.

If $i < \delta$, we might say Mr. Kuchenfresser is relatively impatient. In other words, he discounts at a rate that is higher than what can be earned in the bank. As a result, he would prefer to borrow (if possible), and is even willing to give up some future consumption in order to finance current consumption. As a result, $dc/dt < 0$. In this case, Mr. K's consumption tends to fall over time because he prefers to consume sooner rather than later.

If $i = \delta$ then $dc/dt = 0$. Mr. Kuchenfresser's rate of impatience is matched by the bank's rate of interest, and so, in a relative sense, he is neither patient nor impatient (relative to the market), and neither saves nor borrows. Mr. K's consumption is constant over time.

With a closed form for $U(c)$ this problem can be solved explicitly for Mr. Kuchenfresser's equilibrium consumption and savings paths. Adding an initial capital stock and any restrictions on final wealth allows us to pin down an exact set of values for consumption and savings. Many variations on this basic problem are possible. These variations form the core of dynamic microeconomic theory. They also form the basis of much modern economic growth theory and macroeconomic investigation.

Transversality Conditions

We close this chapter by noting that in many cases in economics, the endpoints of a problem, in particular the terminal point, may not be fixed. In solving problems such as these with variable endpoints, we typically face extra freedom in seeking a solution. For example, the decision maker in the problem may have some choice regarding the final time period (e.g., when to shut down operations or when to pull a product from the market), the final state (e.g., the amount of ore to leave in a mine), or both. In these cases, one or more boundary condition is missing and extra conditions are needed to "pin down" an optimal solution. These extra conditions are referred to as **transversality conditions**. Transversality conditions describe the properties of a dynamic system as it moves across the finish line of the problem. The transversality condition can be viewed as similar to the complementary slackness constraints in a static optimization problem. The transversality conditions must be satisfied, but they need not be binding. Unlike the Euler equation in the calculus of variations, which applies to all points along the optimal path, the transversality condition applies only to one point of time along the optimal path, most commonly the final period of the problem, T. The transversality condition does not replace the Euler equation. It is an additional condition for optimality. A proper definition of a dynamic optimization problem, including those we will encounter as optimal control or dynamic programming problems, will always include either a clear specification of endpoints or the transversality conditions that replace them. Readers interested in the subtleties of the transversality conditions, especially as they apply to infinite-horizon problems, can consult (Kamihigashi 2001) and the references cited therein.

References

Benveniste, L. M., & Scheinkman, J. A. (1982). Duality theory for dynamic optimization models of economics: The continuous time case. *Journal of Economic Theory, 27*(1), 1–19.

Chiang, A. C. (2000). *Elements of dynamic optimization*. Waveland Press.

Kamien, M. I., & Schwartz, N. L. (1991). *Dynamic optimization: The calculus of variations and optimal control in economics and management* (2nd ed.). North-Holland Press.

Kamihigashi, T. (2001). Necessity of transversality conditions for infinite horizon problems. *Econometrica, 69*(4), 995–1012.

Equations of Motion

6

In most cases, the essential problem associated with solving a dynamic optimization puzzle is to describe the process of change in variables of interest over time. In Chap. 2, it was pointed out that both difference equations and differential equations can be used to describe changes in variables over time. Continuous-time problems, as in the Euler equation example of Chap. 5, rely on differential equations. Discrete-time problems, such as those of Mr. Kuchenfresser's in earlier chapters, use difference equations. Because these equations are central to dynamic optimization problems, you are likely to encounter them in future studies. Therefore, in this chapter, we explore them in a bit more detail. The goal here is not to be comprehensive. The hope is that a bit of prior exposure will help when you encounter these topics in other settings.

Time is the Key

We encountered equations of motion implicitly in Chap. 1, where we used simple accounting to track changes in the size of our party cake. Then, in Chap. 2, the concept was introduced more formally as the transition equation linking Mr. Kuchenfresser's state variable across periods. In Chap. 5, we saw how the equation of motion is directly imbedded in the integrand in a calculus of variations problem.

In all of these examples, time has played a central role, as it does in many economic problems. Time may sometimes appear as an independent argument, but more often, time appears because the variables of interest (e.g., the state variable and control variable) are indexed by time. For example, in Chap. 4, consumption and savings for Mr. Kuchenfresser were subscripted by time. And we've already encountered the simplest equation of motion of all time in the form of the cake-eating problem, in which saving is ruled out and the only decision to make is how

© The Author(s), under exclusive license to Springer Nature Switzerland AG 2025
G. Shively, *A Beginner's Guide to Dynamic Optimization in Economics*, Classroom
Companion: Economics,
https://doi.org/10.1007/978-3-032-09374-5_6

to consume a fixed pool of a resource over a known time horizon. The stock evolves (downward only) according to consumption decisions:

$$k_{t+1} = k_t - c_t.$$

In this case, Mr. Kuchenfresser's cake can only get smaller with each slice eaten, and his total consumption is capped by the initial size of the cake. The cake's evolution is defined by a simple difference equation.

Difference Equations

Technically, the fact that the derivative of a function is defined as the limit of tiny discrete differences leads to a direct analogy between the calculus of finite differences, which is used in the context of difference equations, and differential calculus, which is used for differential equations. Being comfortable with both is useful. In practice, many theoretical problems are written in continuous time, but a large number of empirical problems are written in discrete time. However, while difference equations can often be used to approximate differential equations, important and interesting exceptions to the rule do arise. Goldberg (1986) provides a comprehensive discussion of difference equations with numerous examples.

The example of how out-of-equilibrium markets might move toward equilibrium over time is a dynamic, discrete-time problem that has been important in economics since the 1930s. The so-called **cobweb model** (Kaldor, 1934) was developed by the Hungarian economist Nicholas Kaldor based on his observations of agricultural markets (corn and rubber, to be exact). Most agricultural markets, including both crop and livestock markets, are characterized by time lags between production and harvest or sale, time lags between demand and supply, and uncertainty regarding both due to the vagaries of weather and other factors. Lags occur because farmers must make individual planting decisions at the start of the season, based on prevailing prices or expectations of what prices might look like much later in the year when crops are harvested and the aggregate supply from all farmers is realized. In its simplest form, the cobweb model is defined by a pair of difference equations, one for supply and another for demand. The dynamics of the market system can be understood quite easily by observing the interplay of supply and demand in a supply and demand diagram.

To understand how the cobweb model works, imagine a market in which supply and demand are slightly out of equilibrium. Under equilibrium conditions, of course, the market-clearing price would be determined by the intersection of the supply (S) and demand (D) curves in Fig. 6.1. But let's say that because of unfavorable weather during the growing season, farmers experience unexpectedly low levels of output, leading to a one-time backward shift in the supply curve to S_0. In the diagram, we can see that this results in the market-clearing price and quantity combination p_0 and q_0, namely a high price and a low quantity. We have, in essence, defined the momentary (static) equilibrium. But what might happen next?

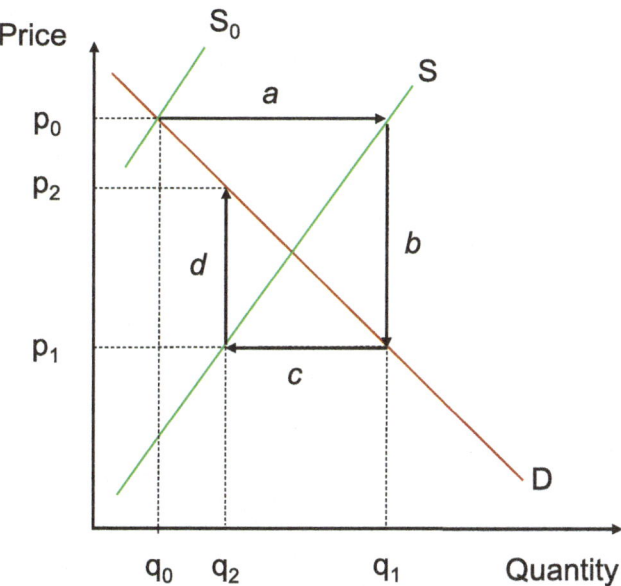

Fig. 6.1 A cobweb model

Kaldor's insight was to recognize that behavior depends on information and expectations. When we add time and behavioral dynamics to the example, we can imagine that farmers, observing a high price in the market, might respond by planting more acres in the next season.[1] The result is that supply increases to q_1 (the rightward shift in the diagram marked *a*) and price falls to p_1 (the downward shift in the diagram marked *b*). The lower subsequent price, p_1, discourages farmers and leads them to produce less (the leftward shift in the diagram marked *c*), leading to q_2. This, in turn, leads to a higher price, p_2, (as indicated by the upward shift in the diagram marked *d*). The back-and-forth continues, and, so long as the supply curve is steeper than the demand curve, iterations over time will lead to **convergence**, in the process tracing out an inward spiral that looks like a cobweb (hence the name). In contrast, if the supply curve is less steep than the demand curve, then the fluctuations will lead to divergence, and equilibrium will never be achieved.

The cobweb model illustrated in Fig. 6.1 can be easily written in terms of a difference equation, especially when we assume linear forms for supply and demand, as we have done. Using Greek letters as slope-intercept parameters, we can write:

---

[1] Price-chasing behavior is so prevalent in some low-income farming communities that farmers will speak of aiming for "jackpot harvests," hoping to harvest a cash crop exactly when scarcity has driven prices higher. Often, since everyone has the same idea and demand for their perishable output is inelastic, the wager ends in a bust.

$$q_t^D = \alpha + \beta p_t$$

$$q_t^S = \gamma + \delta p_{t-1}$$

along with the equilibrium condition:

$$q_t^S = q_t^D.$$

Simple substitution and rearrangement lead to:

$$p_t = \left(\frac{\alpha - \gamma}{\beta}\right) - \left(\frac{\delta}{\beta}\right) p_{t-1}$$

which is a first-order difference equation. In this simple version of the cobweb model, an equilibrium is defined by the **fixed point** where:

$$p_t = p_{t-1}.$$

Let's call this equilibrium point p* since it must hold for all time points. Then it is the case that:

$$p^* = \left(\frac{\alpha - \gamma}{\beta}\right) - \left(\frac{\delta}{\beta}\right) p^*$$

or

$$p^* = \left(\frac{\alpha - \gamma}{\beta + \delta}\right).$$

In other words, the long-run equilibrium price in the market becomes a function of the parameters of the supply and demand schedules, but not time. In technical terms, we say the difference equation is **autonomous**, meaning the solution is independent of time.

Other cases in which the cobweb model might apply can be imagined, and plenty of ink has been spilled by economists in arguing whether the kind of **adaptive expectations** illustrated by the cobweb model are an accurate representations of how economic agents actually form expectations. As an alternative, **rational expectations** are often posited as a more accurate reflection of how people make economic decisions. Producers with rational expectations would understand how markets work and be able to see through the potential mistakes associated with over-reacting to price changes, leading to less erratic price and quantity movements in the marketplace than the cobweb model predicts.

As stated previously, the example illustrated above involves iterations that eventually lead to convergence. The economic requirement for convergence in this particular model is that the supply curve is steeper than the demand curve. In that case, we observe an inward spiral to an equilibrium. Furthermore, displacements from that equilibrium will lead to a return to the equilibrium. The simple dynamics of the system are said to have a **stable equilibrium**. In contrast, if the supply curve is less steep than the demand curve, then displacements from equilibrium lead to divergence and the system never returns to equilibrium. In other words, the system is said to have an **unstable equilibrium**. As you encounter more complex examples of dynamic systems, in particular those characterized by two or more equations of motion, you will need to pick up the tools necessary to examine the stability properties of these systems. In Chap. 10, we'll scratch the surface of some of this material.

Simple inventory and backlog problems can be devised where the state variable is the stock of inventory available. We can even consider the possibility of negative stocks corresponding to inventory backlogs. As a simple example, let's imagine a hypothetical retailer that purchases goods from producers and then turns around and places these items on a shelf to sell to consumers. Let's say the monthly stock on hand of a certain item available to sell is x. The stock of the item for the next month depends on the amount of stock on order from the factory, u_t, and monthly sales to consumers, w_t, where:

$$x_{t+1} = x_t + u_t - w_t.$$

This equation describes how the stock evolves over time, and contains enough essential pieces that we could think about scenarios in which, for some reason, w_t exceeds $x_t + u_t$ and the retailer runs out of the item. In fact, u_t could itself embody supply chain dynamics, and could be subject to shocks or disruptions, as much of the world experienced during the COVID-19 pandemic. From a practical point of view, given enough information about consumer demand in the past, the retailer could even begin to generate forecasts of w (say, on a seasonal basis). The equation of motion could then be imbedded in a problem in which the decision-maker's objective function might be to maximize revenue subject to inventory costs and the possibility of stock-outs due to stochastic demand shocks. If inventory backlogs lead to loss of customers, stock-outs might reduce future revenues, and it could prove optimal to carry extra levels of inventory. In some formulations, the cost of a stock-out is incorporated using a **penalty function**, an idea that can be traced back to a paper by Arrow et al. (1951). Because inventory models are so important and interesting, we'll return to look at a few more of them in the next chapter.

Differential Equations

Recall that a solution to an algebraic equation is a number or set of numbers that will satisfy the equation. A solution to a differential or difference equation is a function that satisfies the equation at all times. If we use the notation t for time and $x(t)$ for a variable whose value is a function of t, then it is conventional to write the time derivative of x as: $dx(t)/dt = \dot{x}(t) = \dot{x}$, where dx/dt is understood to be the total derivative of x with respect to t (sometimes written x', as we did in Chap. 5).

To link the rate of change in x to values of x and t, we use a differential equation $f(x(t), \dot{x}(t), t)$. A solution to a differential equation is a function $x(t)^*$ such that the arguments satisfy the differential equation identically, i.e., $f(x(t)^*, \dot{x}(t)^*, t) = 0$. An **ordinary differential equation** has only one independent variable. A **partial differential equation** has multiple independent variables, and hence partial derivatives appear in the solution. Although, this book is not intended to teach you everything you might need to know about differential equations, a few of the more important cases and examples are included below. A thorough introduction to ordinary differential equations is provided by Ross (1991).

Differential equations can take a number of forms. Here, we review a few of the most basic general forms that appear in the economics literature. We've already encountered some of these forms in Chap. 5 in the context of the calculus of variations.

Example: Depreciating Investment

Consider an investment that *decreases* in value over time at a constant rate, where net investment is understood to be the change in the capital stock:

$$I(t) = k'(t) = \frac{dk}{dt} = 100e^{-0.05t}.$$

This is a **first-order ordinary differential equation (FODE)**. We can solve it directly by integration:

$$\int \frac{dK}{dt} dt = \int 100e^{-0.05t} dt.$$

This gives the so-called **general solution** for the problem, which is:

$$k(t) = \frac{-100}{0.05} e^{-0.05t} + c$$

where c is an arbitrary constant.

A general solution describes the basic pattern of movement of the variable of interest, but cannot pin-down any particular values for the variable. This lack of determinacy stems from the fact that an arbitrary constant, c, in the example above, appears in the solution. To know $k(t)$ exactly at any point along its path, we need to know its value at one particular date, either $t = 0$ or $t = T$. As we have seen, pinning-down the variable in this way describes the system at its endpoint. The solution in this case is referred to as a **particular solution**, since it is specific to the chosen value for the arbitrary constant.

In the example above, to turn the general solution into a particular solution, it is sufficient to specify a starting value for k. With a starting point, we can find the particular solution to the differential equation, i.e., an exact solution. Using $k(0) = 1000$, for example, the equation becomes:

$$1000 = -2000e^0 + c$$

which can be solved to obtain a value for the constant $c = 3000$. Substituting this value back into the general form of the solution gives us the particular solution:

$$k(t) = 3000 - 2000e^{-0.05t}.$$

Note that with a different starting point, we would get a different particular solution. This is because depreciation paths differ according to the level of the initial investment.

The general form for a linear first-order differential equation (FODE) with constant coefficients is:

$$\dot{x} + ax = b, a \neq 0.$$

This form can be solved by multiplying through on both sides by the so-called **integrating factor**, e^{at}. Performing the multiplication results in:

$$e^{at}x = \left(\frac{b}{a}\right)e^{at}.$$

This suggests the general solution:

$$x(t) = \left(\frac{b}{a}\right) + Ce^{-at}.$$

The general form for a linear FODE with a variable coefficient is:

$$\dot{x} + a(t)x = b(t).$$

Here, a different integrating factor must be used, namely:

$$I(t) = \exp\left(\int a(t)dt\right).$$

Integrating gives:

$$I(t)\dot{x} + a(t)I(t)x = I(t).$$

The general solution for this form is:

$$x(t) = \frac{\int I(t)b(t)dt}{I(t)}.$$

Example: First-order Differential Equation (FODE) With Variable Coefficient

Consider the equation:

$$\dot{x} + 4tx = 2t.$$

Here $a(t) = 4t$ and $b(t) = 2t$. The integrating factor is: $I(t) = e^{2t^2}$. Multiplying each side by the integrating factor gives:

$$e^{2t^2}\dot{x} + 4te^{2t^2}x = 2te^{2t^2}.$$

Integrating both sides results in:

$$\int 2te^{2t^2}dt = \frac{1}{2}e^{2t^2}.$$

Solving for $x(t)$ leads to:

$$x(t) = \frac{1}{2} + ce^{-2t^2}.$$

Some differential equations are nonlinear. These are typically more difficult to solve. Sometimes the equation can be solved if the terms in x and t can be separated. In other words, it is "separable" if it can be written as:

$$f(x)\dot{x} = g(t)$$

in which case we can integrate both sides with respect to t:

$$\int f(x)\frac{dx}{dt}\,dt = \int g(t)dt \quad \text{or} \quad \int f(x)dx = \int g(t)dt.$$

Example: A Nonlinear FODE

As an example, we can solve $3x^2\dot{x} = 4t^3$. We have:

$$\int 3x^2\dot{x}\,dt = \int 4t^3\,dt$$

$$x^3 = t^4 + c$$

$$x(t) = \left[t^4 + c\right]^{1/3}.$$

We will encounter another nonlinear FODE in this chapter's appendix.

Qualitative Solutions

Typically, economic problems are such that we cannot obtain analytical solutions to the differential equations of interest. In these cases, it still may be possible to learn something about the behavior of a dynamic system using **qualitative analysis**. Qualitative analysis is usually carried out through the use of a **phase diagram**, which is not unlike a weather map that shows the direction and strength of wind patterns. A phase diagram gives a graphical description of the dynamics of a variable or pair of variables. Here, we provide a basic introduction. Subsequent chapters address more complex cases, as does Shone (1997).

In the case of a single variable problem, values of the variable of interest are typically plotted against the variable's time derivative. In multiple variable problems, the labeling of axis for plotting will depend on the issue of interest.

To illustrate the use of a single-variable phase diagram, consider a problem in which the output of an economy at time t depends only on the capital stock at time t. Capital depreciates at the rate δ and can be accumulated through savings. The parallels and complications to Mr. Kuchenfresser's consumption-savings problem should be apparent. We define variables as follows:

$k(t)$ capital stock at time t

$q = f(k)$ output, produced via the concave production function $f(k)$, with $f(k) > 0$, $f(k) < 0$

s savings (assumed to be a constant share of output)
$I = sf(k)$ investment
$0 < \delta < 1$ capital stock depreciation.

Defined in this way, the growth in the capital stock can be written as:

$$g(k) = sf(k) - \delta k.$$

In words, this simply says that growth in the capital stock depends on savings from production minus depreciation of the stock. To develop some intuition and insight, we can graph of $g(k)$. It is conventional to place the stock, k, on the x axis and its rate of growth, $\dot{k}$, on the y axis, which has been done in Fig. 6.2. This graph is known as a **phase line**. It describes the underlying dynamics of how k evolves.

Figure 6.2 indicates that at low levels of the capital stock, the rate of capital accumulation, $\dot{k}$, is positive and rising. As a result, the capital stock itself is growing. At higher levels of the capital stock, the growth rate of the capital stock is positive but falling. In this range (below k_1 in the graph), the capital stock continues to grow but at a lower rate. These patterns result from the concavity of the production function. At low levels of capital, the incremental boost from a bit more capital is larger. Eventually, however, capital becomes less productive. Beyond some point (in the graph, k_1) large capital stocks are accompanied by negative growth rates in capital and a falling capital stock. What is the logic behind this phase diagram?

First, assume $sb - \delta > 0$, i.e., the rate of savings exceeds the rate of depreciation. This suggests that the rate of capital accumulation can be zero only at two points: either where $sf(k) = \delta k$ or where $k = sq/\delta$. This implies two equilibrium

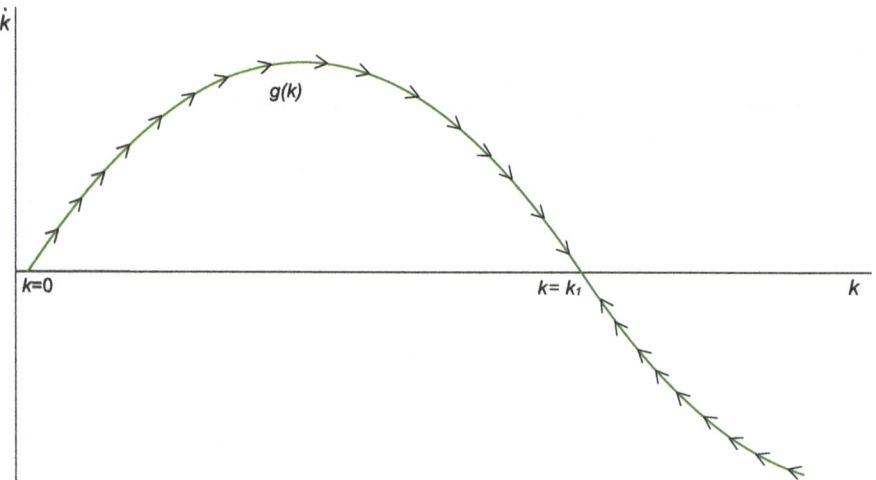

Fig. 6.2 Phase line for changes in a capital stock

points: one where $k = 0$ and another where $k = k_1$. Two cases highlight the dynamics of this simple dynamic system.

Case 1. $0 < k < k_1$. The capital stock is relatively small, in which case $g(k) = sf(k) - \delta k > 0$. This implies growth is positive $(\dot{k} > 0)$ and k is rising.

Case 2. $k_1 < k$. The capital stock is relatively large, in which case $g(k) = sf(k) - \delta k < 0$. This implies growth is negative $(\dot{k} < 0)$ and k is falling.

Renewable Natural Resources

Many dynamic problems that relate to the use of renewable natural resources rely on specific equations to measure the growth of a resource stock over time. For example, consider the equation of motion for a population of fish in a lake. The resource is renewable, meaning it can grow on its own under the right conditions, but it is also potentially exhaustible, meaning that if the size of the population falls to zero (or some critically low level), then it cannot recover from this collapse.[2] One of the most common functional forms for the growth equation of a biological resource stock is the logistic model, in which the population increases slowly at low population levels, then more rapidly, then more slowly again. When written as a differential equation, the logistic model takes the form:

$$\frac{dX(t)}{dt} = rX(t)\left(1 - \frac{X(t)}{K}\right)$$

where X is the size of the stock, r is the rate of growth of the stock, and K is the **carrying capacity** of the lake. What the equation tells us is that when the stock of fish is low, the total addition to the stock, $rX(t)$, is also low. As the stock gets larger, the number of new fish added to the stock increases. Then, as the stock approaches the carrying capacity of the lake, growth slows again. Why might we see such a pattern in nature? Ecologists have found that growth may be low when the population is small due to a low probability of breeding success. Imagine a small number of fish swimming around in a large lake and not finding each other. (Unfortunately, there are no dating apps for fish). If the population is slightly larger, chance encounters leading to breeding success become more likely. But then, when the population grows exceedingly large (approaching K in the model), reproductive success falls because all of the existing fish are competing for food and necessary habitat in which to lay their eggs. When graphed, this relationship between stock and growth looks

---

[2] Technically, we say that the state from which the population cannot recover is an **absorbing state**. For some species subject to **critical depensation** this population size may be greater than a single reproductive pair, which raises important issues for species protection. We'll dig a bit more deeply into absorbing states and related concepts in later chapters.

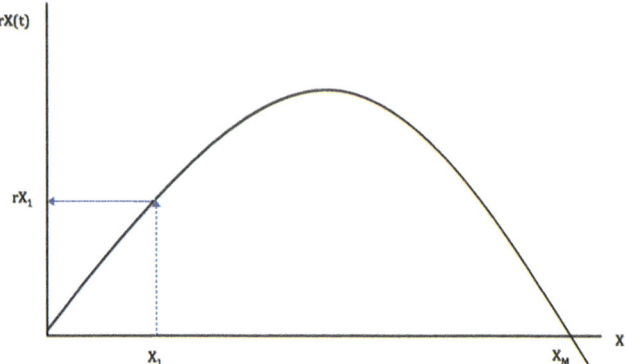

Fig. 6.3 The growth of a renewable natural resource

something like the pattern of growth illustrated in Fig. 6.3. The total size of
the stock is measured on the horizontal axis, and the growth of the stock,
$rX(t)$, is measured on the vertical axis. When X is low, $rX(t)$ is low. Similarly,
on the right side of the graph, when X is large, $rX(t)$ is also low. In between,
there is a point where the population reaches its maximum rate of population
growth. In theory, at that point, as long as each period's harvest doesn't
exceed rX, that level of harvest could be maintained in perpetuity, since in
each period, the stock would continue to replace the amount harvested in the
previous period.

How does time figure in all of this? You can imagine various points on the graph
as corresponding to steps in time. For example, starting from the left side of the
diagram, a specific level of stock at a point in time (indicated by X_1), leads to
a particular increment in the stock. This increment, rX, can be read off the y axis. In
the next period, the stock grows by the increment rX, and so on, marching up and to
the right along the curve. Beyond peak growth (where rX is at its maximum and we
reach the top of the parabola), larger stock values produce growth, but the increment is smaller than before due to, say, competition for food and places to hide
from predators. At X_M, which represents the maximum carrying capacity for the
population, no further growth in the population can occur. To make this point clear,
imagine a scenario on the right-hand side of the graph in which someone tries to
temporarily push the resource beyond this maximum level, say by stocking finger-
lings into a lake that is already at capacity. With insufficient food and habitat, the
result would be a temporarily negative rate of growth, as fish compete, some of
them die, and the population falls back to X_M.

This example is meant to illustrate the practical application of a differential
equation to a real-world issue. If you are interested in this branch of natural
resource economics, you can find many examples. Rigorous treatment of the
topic can be found in Clark (1990). In the appendix to this chapter, we push this
canonical example a bit further and explore the model under various values for r
and X_M, and different initial values for X.

Brownian Motion

Let's now examine a somewhat different equation of motion, which you may encounter if you pursue an interest in finance. A popular and often-used equation of motion to track the evolution in the prices of speculative assets, such as stocks, is so-called **Brownian motion**.[3] Brownian motion characterizes a stochastic process that seems to follow no particular direction—sometimes called a **random walk**. Brownian motion with drift, on the other hand, describes a process which remains random but tends to move in one direction or the other, as in:

$$dx = \alpha dt + \sigma dz$$

where α is a parameter describing drift (an increase up or down) and σ is a parameter that measures the variability in the stochastic process. In some cases, such stochastic processes might exhibit **mean-reversion**, meaning that values tend to drift up or down over time but eventually return to a predictable level. Such a process can be described as evolving according to:

$$dx = \theta(\bar{x} - x)dt + \sigma dz.$$

Some observers have argued that, historically at least, oil prices have followed a mean-reverting process. Dixit and Pindyck (1994) provide a number of applications of such equations of motion to problems in economics and finance. Note that in general, solving stochastic differential equations such as these requires the use of stochastic calculus, also called Itô calculus, after the Japanese mathematician who first described it. See Karlin and Taylor (1975) for details.

Solving a Simple Fisheries Model

To close the chapter, let's return to our example of harvesting fish from a lake. Recall that the equation of motion for the fish population was presented as:

$$\frac{dX(t)}{dt} = rX(t)\left(1 - \frac{X(t)}{K}\right).$$

---

[3] Like many concepts in economics, this one was borrowed from physics. Brownian motion is named after the botanist Robert Brown who, in 1827, observed and described the erratic behavior of pollen in water. In 1905, a young Albert Einstein published a paper that provided a mathematical model of Brownian motion to explain the movement of atoms. The economists Robert Merton and Paul Samuelson are generally credited with transferring the concept to asset markets, where it often seems (especially to those who have tried unsuccessfully to beat the stock market) as if the prices of risky assets behave as if they were grains of pollen in water.

The solution to this differential equation is a bit tedious to derive, but is a good example of using the method of separation of variables. Start with the equation above and separate terms in x and t to obtain:

$$\frac{dx}{x\left(1 - \frac{x}{K}\right)} = rdt$$

or

$$\frac{-K}{x(x - K)}dx = rdt.$$

We can integrate both sides, which results in:

$$ln\,x - ln(K - x) = rt + c$$

or

$$ln\left[\frac{x}{K - x}\right] = rt + c.$$

This can be rewritten as:

$$\frac{x}{K - x} = ce^{rt}.$$

Simplifying yields:

$$x = \frac{Kce^{rt}}{1 + ce^{rt}}$$

or

$$x = \frac{K}{1 + ce^{-rt}}.$$

The constant c, of course, is arbitrary and could be made definite via a starting population, if we knew it. To understand what growth of the population looks like, a graph of $x(t)$ against time, using the parameter values $K = 500$ and $r = 0.20$ looks like Fig. 6.4.

To turn this biological model into an economic problem, we need to add a few things, in particular, the cost of fishing effort and the value of harvests. With harvesting, the rate of change in the resource stock must reflect natural growth *and* offtake. If harvest $H(E, X)$ takes the form qEX then sustained yield

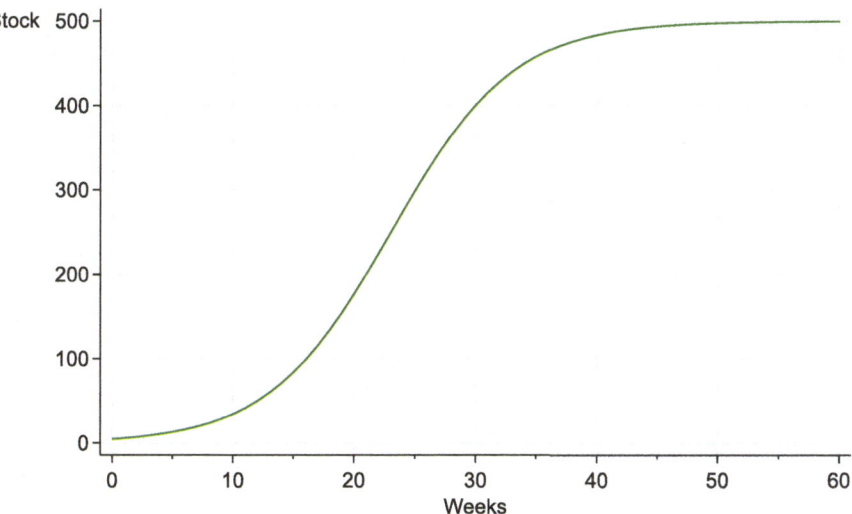

Fig. 6.4 Logistic resource growth over time

(the amount of annual offtake that is exactly compensated by growth in the stock) implies:

$$\dot{X} = F(X) - Y = 0.$$

In words, the level of stock is constant (i.e., growth equals harvest). Recall we assume:

$$Y = H(E, X) = qEX.$$

Therefore, for $F(X) = rX(1 - X/K)$ we have:

$$Y = rX\left(1 - \frac{X}{K}\right) = qEX.$$

Solving for X gives:

$$X = K\left(1 - \frac{qE}{r}\right)$$

and yield becomes:

$$Y = qEX = qKE\left(1 - \frac{qE}{r}\right).$$

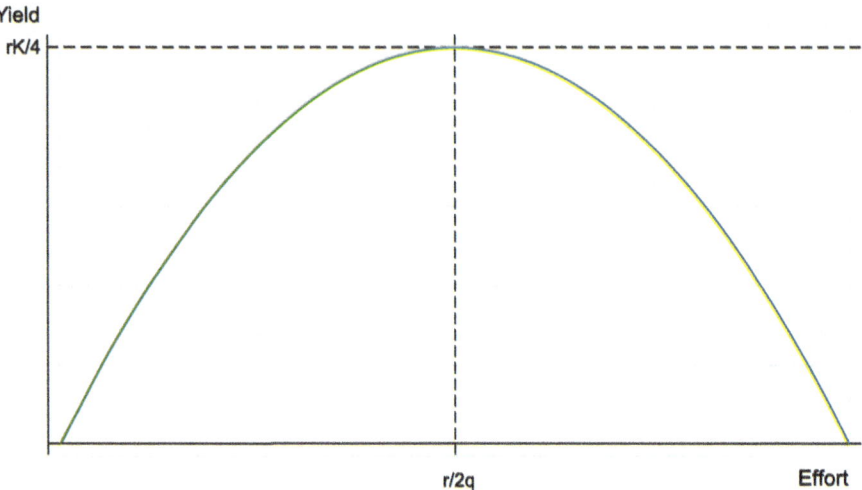

Fig. 6.5   A yield-effort curve for a renewable resource

This yield function is graphed in Fig. 6.5. Like what we saw before in Fig. 6.3, this function is a parabola, in this case with a maximum yield at $Y = rK/4$, corresponding to maximum effort at $E = r/2q$.

Note that, except under very special circumstances, maximum yield at maximum effort will *not* be the economically optimal harvesting decision that maximizes the economic return from the resource. Instead, when fishing effort is costly (in terms of time or equipment), the economic goal will be to equate the marginal cost of fishing (e.g., for one more hour) with the marginal benefit of fishing (e.g., for one more fish). As a result, optimal effort will occur to the left of $r/2q$.

If we go a bit further with our current example and assume some values for parameters, say $r = 0.5$, $K = 10$, and $q = 0.1$, we can write an explicit form for yield, namely:

$$Y = H(E) = (0.1)(10)E(1 - 0.2E).$$

or

$$Y(E) = E - 0.2E^2.$$

Note that in this example, $Y = 0$ at $E = r/q$, i.e., at $E = 5$. If $E > r/q$, i.e., if effort is sufficiently high, yield will be zero. If the relative rate of harvest qE is greater than the natural rate of population growth r, then the stock will decline until the population is driven to extinction. Examples of such a resource might include anything that reproduces slowly but commands a high price, which encourages rapid exploitation. Whales, elephants, and redwoods come to mind.

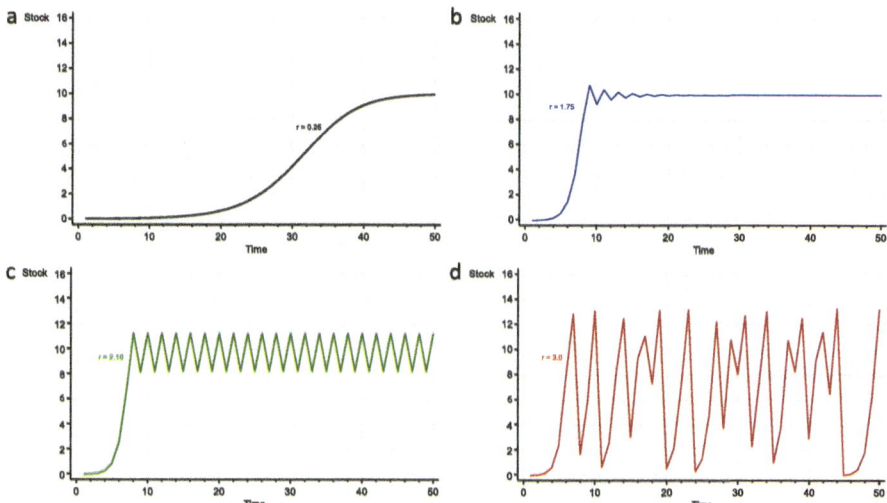

Fig. 6.6 Resource stock growth under four different growth rates

Not surprisingly, sustainable use of such resources often requires strict enforcement of the rules regarding harvest from these stocks.

Now that we have a solution for the differential equation, we can generate graphs of the stock against time. These can be easily constructed forward in time using a spreadsheet and the corresponding difference equation. Panels (a)–(d) in Fig. 6.6 provide four examples. With $r = 0.25$, panel (a) of Fig. 6.6 shows a gradual approach to the carrying capacity. With $r = 1.75$, panel (b) exhibits damped oscillation converging to the carrying capacity. With $r = 2.10$, panel (c) indicates periodic behavior around (but not converging to) the carrying capacity. Finally, with $r = 3.00$, panel (d) illustrates **deterministic chaos**. These results suggest the dynamic behavior in a discrete-time setting can vary widely (and wildly) depending on minor variations in the growth rate of the stock. Conrad and Clark (1987) provide additional details and insights into the implications of this model.

Example: Harvesting Rules in the Real World

Imagine for a moment that you have been asked to propose a harvesting rule for fish from the lake described above, but that the rate of growth in the fish population is uncertain. What kind of harvesting rule might you propose? As the previous analysis shows, harvesting a natural resource with only incomplete knowledge of the true rate of growth could be problematic because too great a harvest could drive the stock to zero. Although our original problem was not cast in terms of uncertainty, we can easily see that without clear knowledge of r, we would be operating in an uncertain world. If nothing is known about the true rate of growth of the resource, it might be necessary to sketch out possible rates of growth along with their probabilities and work from that. Or, it might be possible to learn something about r by cautious harvesting and observation of the stock. A prudent harvesting rule that permits some **learning-by-doing** might be useful. The danger is that, if the stock requires a positive minimum viable population, harvesting heavily from a stock that might be "crashing" could spell disaster. More than anything, this example illustrates the need for a strong ecological understanding of how natural systems work before exploiting them for economic gain. It also suggests that finding the solution to tricky problems like this requires that economists and biologists work together.

References

Arrow, K. J., Harris, T., & Marschak, J. (1951). Optimal inventory policy. *Econometrica, 19*(3), 250–272.

Clark, C. W. (1990). *Mathematical bioeconomics: The optimal management of renewable resources*. Wiley.

Conrad, J. M., & Clark, C. W. (1987). *Natural resource economics: Notes and problems*. Cambridge University Press.

Dixit, A., & Pindyck, R. (1994). *Investment under uncertainty*. Princeton University Press.

Goldberg, S. (1986). *Introduction to difference equations with illustrative examples from economics, psychology, and sociology*. Dover Publications.

Kaldor, N. (1934). A Classificatory note on the determination of equilibrium. *Review of Economic Studies, 1*, 122–136.

Karlin, S., & Taylor, H. M. (1975). *A first course in stochastic processes*. Academic Press.

Ross, S. L. (1991). *Introduction to ordinary differential equations* (4th ed.). Wiley.

Shone, R. (1997). *Economic dynamics. Phase diagrams and their economic application*. Cambridge University Press.

Inventory and Stock Adjustment Models

7

Intertemporal problems of the firm, especially those involving production planning and inventory control, are natural candidates for analysis using the methods of dynamic optimization. Production and inventory problems typically share two key features: an objective function that expresses the firm's desire to either maximize profits from sales or minimize the cost of meeting a production and inventory target; and a financial tradeoff between the cost of producing to meet contemporaneous demand and the cost of storage to meet future demand. Accordingly, and not surprisingly, in such problems the solution is one in which firms must balance two marginal costs: the marginal cost of producing one unit of output in the current period and storing it for later use; and the marginal cost of producing one unit of output at a future date. As mentioned in the last chapter, the formal analysis of inventory problems by economists can be traced to the early work of Arrow et al. (1951). Since then, numerous applications have appeared in both the microeconomics and the macroeconomics literature. The most interesting inventory and stock adjustment problems are those in which some aspect of the problem, most often demand, is treated as stochastic. Before investigating the characteristics of stochastic problems, we first examine some deterministic examples.

Deterministic Models of Stock Adjustment

We focus in this chapter on a pair of simple deterministic problems of production and inventory control. The first of these was originally outlined by Kamien and Schwartz (1991). The problem illustrates the way in which simple constraints can be imposed along the optimal path. The problem demonstrates the basic variational calculus approach to setting up and solving inventory problems.

© The Author(s), under exclusive license to Springer Nature Switzerland AG 2025
G. Shively, *A Beginner's Guide to Dynamic Optimization in Economics*, Classroom Companion: Economics,
https://doi.org/10.1007/978-3-032-09374-5_7

Example: Production and Inventory Control to Meet a Target

Consider a firm that must plan production and inventory to minimize the sum of production costs and inventory holding costs. Let's identify inventory as x. This means the change in inventory at any point in time, (i.e., production), is dx/dt or $x'(t)$. Let's further define c_1 and c_2 as marginal costs: c_1 is the per unit cost of production, and c_2 is the per unit cost of storage. The cost of holding inventory at any point in time is therefore $c_2 x(t)$ and the cost of producing another unit, which we assume to be quadratic in output, is $c_1 x'(t)^2$. Mathematical intuition should lead you to see that introducing a bit of curvature to the cost function in this way means that solving the problem through production alone is unlikely to be optimal, because as the level of production rises, production cost rises exponentially. The problem, therefore, is formulated in such a way that there is always some potential incentive to store.

Let's imagine the manufacturer has received an order for M items, and that these M items must be delivered by time T. Further, assume the firm currently holds no stock of the items on hand. For simplicity let's suppose the firm does not discount. How should the firm plan production in order to minimize the cost of supplying M items by time T? Even before solving the problem, we can intuit that the firm needs to balance the cost of production and the cost of storage: producing everything now will be costly because the stock will need to be stored until delivery; but while postponing production until the last minute will avoid storage costs, it will likely be prohibitive due to the upward sloping cost of production.

To solve this problem, let $x(t)$ represent the inventory on hand at time t. Since the firm begins with no stock on hand, we have the initial condition $x(0) = 0$ paired with the terminal condition $x(T) = M$. The inventory at any point in time is equal to accumulated past production. The rate of change in the firm's inventory is equal to the rate of production. As alluded to earlier, let us call this rate of production $x'(t)$. If, as above, we define c_1 and c_2 as nonnegative constants governing the marginal costs of production and storage, then the firm's total cost at any point in time is described by:

$$c_1 (x'(t))^2 + c_2 x(t)$$

where the first term corresponds to the cost of production and the second term corresponds to the cost of storage. The firm's problem is to find the production rate $x'(t)$ that minimizes the cost of meeting the production target:

$$\text{Min} \int_0^T \left[c_1 (x'(t))^2 + c_2 x(t) \right] dt$$

subject to $x(0) = 0$, $x(T) = M$, and $x'(t) \geq 0$.

Note that this problem has fixed endpoints. The solution to the problem can be found via a straightforward application of the calculus of variations, provided we are willing to assume that the nonnegativity condition $x'(t) \geq 0$ is never binding. The integrand provides the necessary pieces for constructing the Euler equation. Recalling Eq. (5.4), the Euler equation from Chap. 5, we have $F_x = c_2$, $F_{x'} = 2c_1 x'$, and therefore the Euler equation can be expressed as:

$$x''(t) = \frac{c_2}{2c_1}.$$

Integrating once yields:

$$x'(t) = \frac{c_2 t}{2c_1} + k_1.$$

And integrating a second time yields:

$$x(t) = \frac{c_2 t^2}{4c_1} + k_1 t + k_2.$$

The constants of integration, k_1 and k_2 can be recovered via the firm's boundary conditions, namely $x(0) = 0$ and $x(T) = M$. By simple substitution, we obtain $x(0) = 0 = k_2$ and $x(T) = M = \frac{c_2 T^2}{4c_1} + k_1 T + k_2$. This implies $k_1 = \frac{M}{T} - \frac{c_2 T}{4c_1}$ and $k_2 = 0$. The final form for the optimal inventory path $x(t)$ is therefore given by:

$$x(t) = \frac{c_2 t(t - T)}{4c_1} + \frac{Mt}{T}.$$

For this problem, Kamien and Schwartz (1991) provide an economic interpretation of the Euler equation. The Euler equation requires that we balance the rate of change of the marginal production cost against the marginal cost of holding inventory. In other words, the marginal cost of producing a unit at time t and holding it for an increment of time Δ must be the same as the marginal cost of waiting and producing the unit at time $t + \Delta$. If marginal inventory cost is less than the rate of change in marginal production cost, then it will be optimal to produce more now and store it for later sale. In other words, along the optimal path, we must be indifferent between producing an extra unit and storing it, and avoiding inventory costs by postponing production to a time at which it will be slightly more costly to produce.

Also note that upon inspection of the Euler equation, it is obvious that $x'' > 0$. So, we know x' (production) is increasing over time. Therefore, as long as $x_T \geq c_2 T(T - 2t)/4c_1$, the firm can satisfy the constraint that requires production to remain non-negative along the planning horizon. That is, the constraint will be satisfied as long as production is large relative to the planning horizon or as long as storage cost is low relative to production cost.

Labor Input and Adjustment

This second example is a much-simplified version of a model presented by Hamermesh (1989). The setup is relatively straightforward and illustrates the method of solving a reduced-form homogeneous second-order differential equation via the use of characteristic roots.

A firm requires labor input for production and has decided to increase its labor stock from the current level, L_0, to a target level L_T by a pre-specified time. Adjusting the labor force is costly. The firm must choose the speed of labor force adjustment. For simplicity, we ignore other inputs and assume profit is a concave function of labor input, i.e., $\pi'(L) > 0$ and $\pi''(L) < 0$. The cost of adjusting labor is $c(L') = bL'^2 + k$, where $b > 0$ and $k > 0$. Profit net of the cost of adjusting labor is therefore $\pi(L) - c(L')$. The firm discounts future profits at the rate ρ. The problem for the firm is to maximize total discounted net profit over time while adjusting the size of the workforce to its target level.

In Hamermesh's original statement of the problem, finding a solution is complicated by the fact that T is a choice variable for the firm and the firm has a capitalized value (what we called a salvage value in Chap. 1) in the post-T period that itself depends on the choice of T. Here, we make the simplifying assumptions that T is fixed and that after T, the firm goes out of business. Our problem can be written as:

$$\text{Max} \int_0^T \left[\pi(L) - bL'^2 - k\right]e^{-\rho t} dt$$

subject to $L(0) = L_0$ and $L(T) = L_T$. To find the solution, we use Euler's equation to build the relevant pieces, which are:

$$F_L = \pi'(L)e^{-\rho t}$$

and

$$F_{L'} = -2bL'e^{-\rho t}.$$

Assembling the Euler equation, we obtain:

$$L'' - \rho L' + \frac{\pi'}{2b} = 0.$$

Proceeding further in obtaining a solution requires the use of the terminal points and the specification of a functional form for π. Let us assume that the profit function takes the form $\pi(L) = 2L - dL^2$, with $\rho = 0.95$, $b = 0.5$, and $d = 0.001$. Then we have, $\pi' = 2 - 2dL$. The reduced form of the second-order differential equation for $L(t)$ is given by:

$$L'' - 0.95L' - 0.002L - 2 = 0.$$

Getting to a solution requires a bit of work. First, note that the characteristic equation associated with a second-order homogeneous differential is given by $r^2 + Ar + B = 0$. Provided $A^2 > 4B$, the roots of this characteristic equation (r_1, r_2) are real and distinct, and the general solution to the differential equation takes the form $L(t) = c_1 e^{r_1 t} + c_2 e^{r_2 t}$. Since $A = -0.95$ and $B = -0.002$ in this example, the condition $A^2 > 4B$ is satisfied. The roots can be found via the quadratic formula:

$$(r_1, r_2) = -\frac{A}{2} \pm \frac{\sqrt{A^2 - 4B}}{2}.$$

Making the necessary substitutions, we obtain $r_1 \approx 0.952$ and $r_2 \approx -0.002$. If we further assign some endpoints for the labor stock, say $L(0) = 100$ and $L(5) = 1000$, then we obtain the pair of equations:

$$100 = L(0) = c_1 e^{(0.952)(0)} + c_2 e^{(-0.002)(0)}$$

$$1000 = L(5) = c_1 e^{(0.952)(5)} + c_2 e^{(-0.002)(5)}.$$

These can be solved to obtain $c_1 \approx 7.7836$ and $c_2 \approx 92.2164$. So, the optimal path for labor adjustment is:

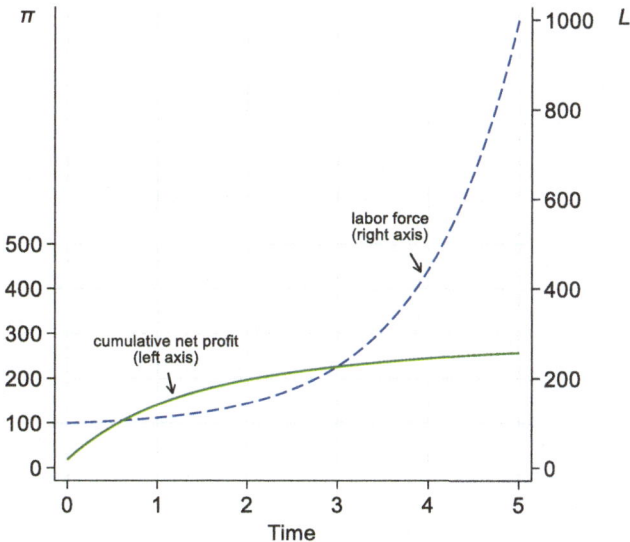

Fig. 7.1 Labor force adjustment over time

$$L(t) = 7.78362 \; e^{0.952t} + 92.2164 \; e^{-0.002t}.$$

Figure 7.1 illustrates the time paths for the labor stock and cumulative net profits

An Application to Vaccines[1]

For another example of inventory control, consider a hypothetical rural health center that anticipates the seasonal outbreak of a common communicable disease. In order to be ready to vaccinate vulnerable individuals in the clinic's catchment area, it wants to increase its stock of the vaccine from the current level of 0 to 5000 doses within ten months. Two vaccines are available. An older version, which is immediately available at zero cost from the central health authority, must be refrigerated, resulting in a cost of s per unit to store. A second, newer version must be stored but does not require refrigeration, and hence represents potential savings in vaccine storage costs of $0.50/ dose. The vaccines are equally effective. Unfortunately, however, production levels for the new vaccine remain low, and global demand exceeds global supply. As a result, the vaccine manufacturer has set the cost for procuring the new vaccine as $c(v) = 0.10v' + 0.01 \left(v'^2 \right)$, where v' is the number of doses of the new vaccine delivered at time t. In other words, faster delivery costs more, and the cost rises exponentially with the speed of delivery. Let's suppose the health agency wishes to minimize the total cost of stocking the vaccine over time, while adjusting the size of the vaccine stock over time to reach its optimal level. Can we find the optimal procurement and storage path for the health agency?

To start, let's consider the production cost for the new vaccine. This is:

$$c[v(t)] = 0.10v'(t) + 0.01v'(t)^2.$$

The per unit savings associated with storage of the (equally-effective) old vaccine is -0.50. As a result, the cost of procuring the new vaccine, if the clinic chooses to order it, is offset by the savings associated with storage costs. Putting these pieces together, we can pose the clinic's cost minimization problem as a calculus of variations problem:

---

[1] Starting from as early as the fifteenth century, people in different parts of the world sought to prevent sickness by intentionally exposing healthy people to smallpox. According to the World Health Organization, this early form of vaccination was brought to Europe in 1721 when Mary Wortley Montagu asked that her daughters be inoculated against smallpox, a practice that she had observed in Turkey. This launched a wave of experimentation that resulted in English physician Edward Jenner inoculating 8-year-old James Phipps (the son of his gardener!) with cowpox in 1796. The boy survived in good health and is generally recognized as the first human to be vaccinated. As a result, we now use the term vaccine, which is taken from *vacca*, the Latin word for cow. Few measures have been as impactful in preventing disease as vaccination: Wahl and Pitzer (2024) estimate that vaccines have saved 154 million lives globally since 1974.

$$\text{Min} \int_0^{10} [(s - 0.5)v(t) + 0.10v'(t) + 0.01v'(t)^2]dt$$

where $v'(t)$ is simply the number of doses delivered at time t and $v(t)$ is the total number of doses in storage at time t. Referring back to the steps outlined in Chap. 5, we can assemble the relevant pieces of the Euler equation:

The objective function is:

$$F(v, v', t) = (s - 0.5)v(t) + 0.10v'(t) + 0.01v'(t)^2$$

which implies:

$$F_v = s - 0.5$$

$$F_{v'} = 0.10 + 0.02v'$$

$$F_{vv'} = 0$$

$$F_{v'v'} = 0.02$$

and

$$F_{v't} = 0.$$

Recalling Eq. (5.6), the expanded form of the Euler equation is:

$$0 = F_{v't} + F_{vv'}v' + F_{v'} + F_{v'v'}v'' - F_v.$$

In this case, we have:

$$0 = 0 + 0 \cdot v' + 0.02v'' - (s - 0.50)$$

or

$$v''(t) = 50s - 25.$$

Integrating once results in:

$$v'(t) = (50s - 25)t + c_1$$

and integrating a second time results in:

$$v(t) = (25s - 12.5)t^2 + c_1 t + c_2.$$

From $v(0) = 0$, $c_2 = 0$. From $v(10) = 5000$, we can calculate:

$$v(10) = 5000 = (25s - 12.5)(10)^2 + c_1(10)$$

or $c_1 = 625 - 250s$. Therefore, the time path for the vaccine stock is:

$$v^*(t) = (25s - 12.5)t^2 + (625 - 250s)t$$

and the time path for procurement is

$$v'(t) = (50s - 25)t + 625 - 250s.$$

Clearly, the actual time path of procurement and storage will depend on the value of s. It is easy to verify numerically that higher storage costs tend to shift production to later in the planning horizon and lower storage costs favor earlier procurement, which should align with your intuition. It is even possible to imagine a scenario in which the exact needs of the clinic are uncertain because the probability, timing, and magnitude of a disease outbreak are uncertain. In that case, a prudent approach might be to build stocks sooner rather than later and to store somewhat more than the expected need to avoid a stockout. Once again, the case for working toward a solution as a multidisciplinary team, in this case consisting of economists, epidemiologists, and public health specialists seems advisable.

References

Arrow, K. J., Harris, T., & Marschak, J. (1951). Optimal inventory policy. *Econometrica, 19*(3), 250–272.

Hamermesh, D. S. (1989). Labor demand and the structure of adjustment costs. *American Economic Review, 79*(4), 684–689.

Kamien, M. I., & Schwartz, N. L. (1991). *Dynamic optimization: The calculus of variations and optimal control in economics and management* (2nd ed.). North-Holland Press.

Wahl, B., & Pitzer, V. E. (2024). Expanded programme on immunization at 50 years: Its legacy and future. *The Lancet, 403*(10441), 2265–2267.

Optimal Control

8

The theory of optimal control is closely related to the calculus of variations, but was developed much later. It is typically associated with the Russian mathematician Pontryagin.[1] The essential characteristic of optimal control theory is the maximum principle. As noted earlier, as an approach to dynamic optimization, optimal control is somewhat more general than the calculus of variations. It was developed in response to some of the shortcomings and restrictions associated with variational calculus. The main advantage of optimal control vis-à-vis the calculus of variations is that it more easily accommodates constraints along the optimal path and hence can be applied to a wider class of problems than the calculus of variations.

The main shift in logic in moving from calculus of variations to optimal control is a new emphasis on an explicit control variable in addition to the state variable of the calculus of variations. As in the calculus of variations, equations describing the rate of change in the state variable are differential equations that relate current choices and values of the state variable to future values of the state variable. In this sense, Chap. 6 provided some of the required foundations regarding equations of motion for the state variable.

Given an initial value for the state variable, a choice of a time path for the control variable leads to a time path for the state variable over the planning horizon. The objective of choosing a control variable path is to maximize (or minimize) some objective function.

---

[1] For those interested in space exploration, Vinter (2010) relates how optimal control emerged as a distinct field of research in the 1950s, and how the relevance of optimal control to the U.S. and Russian space programs was critical to their early successes in launching manned aircraft into orbit.

© The Author(s), under exclusive license to Springer Nature Switzerland AG 2025   79
G. Shively, *A Beginner's Guide to Dynamic Optimization in Economics*, Classroom Companion: Economics,
https://doi.org/10.1007/978-3-032-09374-5_8

The Standard Problem

The standard optimal control problem is to find the control path $u(t)$ that maximizes:

$$V[x(t)] = \int_0^T v(x(t), u(t), t)dt$$

subject to:

$$\frac{dx}{dt} = f(x(t), u(t), t)$$
$$x(0) = x_0$$
$$x(T) = x_T.$$

As written, $x(t)$ is the state variable and $u(t)$ is the control variable. Although endpoints may be free or constrained, just as in a calculus of variations problem, as stated above they are fixed. Optimal control theory allows us to place constraints on the control variable, but for now we assume that $u(t)$ is unconstrained. We also assume that an optimal control exists, is unique, and is differentiable with respect to time. These assumptions are restrictive but can be relaxed.

Necessary Conditions

Necessary conditions for the maximization of an optimal control problem are easily derived after introducing a time-indexed multiplier variable. This variable replaces the Lagrange multiplier used in static constrained optimization problems, as presented in the appendix to Chap. 2. The "dynamic Lagrangian" is the Hamiltonian we encountered in Chap. 5 as Eq. (5.7):

$$H(x(t), u(t), \lambda(t), t) \equiv F(x(t), u(t), t) + \lambda(t)g(x(t), u(t), t) \qquad (8.1)$$

where Eq. (8.1) makes it explicit that each variable has a time path. The **maximum principle** states that the optimal solution to a "standard" Hamiltonian is a triplet that satisfies three conditions.[2]

---

[2] "Standard" in this case means a problem in which derivatives can be used to solve the problem. As we shall see, "must maximize" and $\partial H/\partial u(t) = 0$ are not quite the same thing. The latter, which corresponds to what we are labeling the "standard" Hamiltonian, is more restrictive than the former as it implies the problem is unconstrained and the Hamiltonian is differentiable. Broadly speaking, we only require the control to maximize the Hamiltonian not that a derivative can be taken and set equal to zero.

Condition 1 (The First-Order Condition)

The control variable must maximize the Hamiltonian:

$$\frac{\partial H}{\partial u(t)} = 0. \tag{8.2a}$$

Condition 2 (The State Equation Condition)

The state variable must satisfy a differential equation of the form:

$$\dot{x}(t) = \frac{\partial H}{\partial \lambda(t)}. \tag{8.3a}$$

Condition 3 (The Co-state Equation Condition)

The multiplier (referred to as the co-state variable) must satisfy a differential equation of the form:

$$\dot{\lambda}(t) = -\frac{\partial H}{\partial x(t)}. \tag{8.4a}$$

Boundary conditions also apply. Note that one must exercise a bit of caution when interpreting these conditions. When writing an expression, such as $\partial H / \partial x$ it is understood that the derivative of H is being evaluated at $[x(t),\ u(t), \lambda(t),\ t]$. At each time step, all the conditions for optimality must simultaneously hold. This is the same requirement we placed on Mr. Kuchenfresser when we said it must be the case that, at the optimum, further improvement in the objective could not be realized by reallocating cake from any one period to another.

The three conditions for optimality are sometimes expressed in an expanded form:

$$\frac{\partial F}{\partial u(t)} + \lambda(t)\frac{\partial g}{\partial u(t)} = 0 \tag{8.2b}$$

$$\dot{x}(t) = \frac{\partial H}{\partial \lambda(t)} = g(x) \tag{8.3b}$$

$$\dot{\lambda}(t) = -\frac{\partial F}{\partial x(t)} - \lambda(t)\frac{\partial g}{\partial x(t)}. \tag{8.4b}$$

Note that, unlike calculus of variations problems, continuity is not necessary in optimal control problems. Piecewise-continuous controls are permissible. This allows for jumps in the control variables, which leads to some interesting applications in which rapid switching may be needed. For example, if one is trying to rapidly fill a bathtub with water, the control variable (the faucet controlling the inflow of water) might need to be "fully open" until the tub is full, at which point it switches to "immediately off" in order to avoid overflowing the tub. In this case, the condition described by (8.2a) would not strictly hold, and the control variable would take the discrete values of 0 and 1 with nothing in-between (or, more practically, would be a bounded control with values between 0 and 1).

Complete derivation and proof of the maximum principle is illustrative, but it is also rather tedious to work through and readily available elsewhere,[3] so we will skip it and move straight to an example.

Example: A Problem with One Control Variable

This example is a variation on a problem presented by Leonard and Van Long (1992). It takes the form of a discrete-time maximization problem with one control variable (say, capital) and one state variable (say, consumption). Although, we just managed to work through the general form of a continuous-time version of optimal control, it is instructive to work through a discrete-time problem by hand, as it provides some parallels with examples encountered in earlier chapters.

The problem is to find $\{u_t\}$ to maximize:

$$v = \sum_{t=1}^{3} ln\ (u_t)$$

subject to:

$$x_{t+1} - x_t = 0.25x_t - u_t \qquad t = 1, 2, 3$$
$$x_1 = 1$$
$$x_4 = 1.25.$$

The solution follows six steps.

Step 1: Form the Hamiltonian and optimize it with respect to the control.

The Hamiltonian is:

$$H = ln(u_t) +\ \lambda_t[0.25x_t - u_t]$$

---

[3] Mangasarian (1966) is the standard source for the basic sufficiency theorem for optimal control. Here we assume the second-order conditions hold. For details on the second-order sufficiency conditions see Kamien and Schwartz (1991).

and the first-order condition is:

$$\frac{\partial H}{\partial u_t} = \frac{1}{u_t} - \lambda_t = 0 \qquad t = 1, 2, 3.$$

This defines the optimal control.

Step 2: Find the state equation constraint:

$$x_{t+1} - x_t = \frac{\partial H}{\partial \lambda_t} \qquad t = 1, 2, 3.$$

For this problem, this is:

$$x_{t+1} - x_t = 0.25x_t - u_t \qquad t = 1, 2, 3.$$

Note that this is a first-order difference equation.

Step 3: Find the co-state equation constraint:

$$\lambda_t - \lambda_{t-1} = -\frac{\partial H}{\partial x_t} \qquad t = 2, 3.$$

Note that $\lambda_t - \lambda_{t-1}$ is the change in the shadow value of the resource constraint between periods. Substituting for the current specific problem, this is:

$$\lambda_t - \lambda_{t-1} = -0.25\lambda_t. \qquad t = 2, 3.$$

Again, this is a first-order difference equation.

Step 4: Eliminate the co-state variable using the optimality condition.

We can solve the result from Step 1 for λ. This gives us $\lambda_t = 1/u_t$. We can then substitute this value into the result from Step 3 to obtain $u_t = 1.25u_{t-1}$.

Step 5: Use a boundary condition to determine the initial control.

Using the initial condition $(x_1 = 1)$, the state equation constraint and the result from Step 4, we can solve to get:

$$x_2 = 1.25x_1 - u_1$$

and

$$u_2 = 1.25u_1.$$

Step 6 (actually a series of steps): Substitute recursively.

Solve:

$$x_3 = 1.25x_2 - u_2$$
$$x_4 = 1.25x_3 - u_3$$
$$u_3 = 1.25u_2$$

to get:

$$x_4 = (1.25)(1.25x_2 - u_2) - 1.25u_2$$
$$x_4 = (1.25)^2x_2 - 2(1.25)u_2$$
$$x_4 = (1.25)^2(1.25x_1 - u_1) - 2(1.25)^2u_1$$
$$x_4 = (1.25)^3x_1 - 3(1.25)^2u_1$$
$$1.25 = 1.953125 - 4.6875u_1$$
$$u_1 = 0.15.$$

Now use the solution value for u_1 to recursively obtain:

$$u_2 = (1.25)(0.15) = 0.1875$$
$$u_3 = (1.25)(0.1875) = 0.234375$$
$$u_4 = (1.25)(0.234375) = 0.2929875.$$

Since $x_t = 1/u_t$ we can also find:

$$\lambda_1 = 1/0.15 \approx 6.6666$$
$$\lambda_2 = 1/0.1875 \approx 5.3333$$
$$\lambda_3 = 1/0.234375 \approx 4.2666$$
$$\lambda_4 = 1/0.29296875 \approx 3.4133.$$

Finally,

$$x_1 = 1 \text{(from the initial condition)}$$
$$x_2 = (1.25)(1) - 0.15 = 1.1$$
$$x_3 = (1.25)(1.1) - 0.1875 = 1.1875$$
$$x_4 = (1.25)(1.1875) - 0.234375 = 1.25 \text{ (the endpoint).}$$

So, the entire solution is defined by:

$$\{u\} = \{0.15, 0.1875, 0.234375, 0.29296875\}$$

with:

$$\{\lambda\} \approx \{6.6666, 5.3333, 4.2666, 3.4133\}$$
$$\{x\} \approx \{1, 1.1, 1.1875, 1.25\}.$$

Note that different initial or terminal conditions lead to different optimal paths. Also, for some initial and terminal states, no feasible path will exist.

Savings and consumption revisited

Let's revisit the economic problem from Chap. 4 in which we discovered Mr. Kuchenfresser's time path of per capita consumption to maximize the present discounted value of his stream of utility. Now let's zoom out a bit and think about this problem from a societal point of view. We'll exercise our mental muscles by switching to a continuous time formulation. Define $c(t)$ as consumption at time t, u as utility (a function of c), ρ as a discount rate, $k(t)$ as the ratio of society's stock of capital to its stock of labor at time t, δ as the rate of population growth plus the rate of capital depreciation, and f as the production function. The problem can be stated as:

$$\int_0^T u[c(t)]e^{-\rho t}dt$$

subject to:

$$\dot{k} = f[k(t)] - \delta k(t) - c(t)$$
$$k(0) = k_0 > 0, k(T) = k_T > 0, k(t) \geq 0, c(t) \geq 0$$

where we assume $u' > 0$ and $u'' < 0$ for all $c \geq 0$. We also assume $f(k) < \infty$, $f'(k) > 0$, and $f'' < 0$ for all $k \geq 0$ and that $f(0) = 0$, $f'(0) < \infty$, and $f'(\infty) = 0$.

Let $\lambda(t)$ be the constraint variable for the problem and assume an interior optimum. We can write down the Hamiltonian and state the conditions required for optimality.

The Hamiltonian is:

$$H[k(t), c(t), t, \lambda(t)] = u[c(t)]e^{-\rho t} + \lambda(t)[f[k(t)] - \delta k(t) - c(t)].$$

The choice (control) variable is $c(t)$. The optimality condition is:

$$\frac{\partial H[k(t), c(t), t, \lambda(t)]}{\partial c(t)} = 0$$

which implies:

$$u'[c(t)]e^{-\rho t} = \lambda(t).$$

In words, the first-order condition requires that the optimal path of consumption maximizes the Hamiltonian. We require the shadow value of society's foregone consumption in any current period to be set equal to the discounted marginal utility of consumption in the subsequent period. Using the equation of motion $\dot{k} = f[k(t)] - \delta k(t) - c(t)$ and substituting for $c(t)$, we see that the optimality condition indicates that one forgone unit of consumption is equal to the value an incremental unit of capital. Consuming now displaces capital accumulation and subsequent consumption. At the margin, the optimality condition specifies that society should be indifferent between consuming a unit now and saving it for later, so that everyone can consume more, at a discounted rate, later.

The optimality conditions also include the co-state equation constraint:

$$-\frac{\partial H[k(t), c(t), t, \lambda(t)]}{\partial k(t)} = \dot{\lambda} = -\lambda(t)[f'[k(t)] - \delta]$$

and the state equation constraint:

$$\frac{\partial H[k(t), c(t), t, \lambda(t)]}{\partial \lambda(t)} = \dot{k} = f[k(t)] - \delta k(t) - c(t).$$

The latter defines the evolution of the state variable and indicates that the derivative of the Hamiltonian with respect to the shadow value of capital is equal to a change in the capital stock. When the capital stock changes, the Hamiltonian changes by the amount $\partial H/\partial \lambda$.

The co-state equation indicates how the shadow value of the state variable changes over time. In this problem, the per-period change in the shadow value of the resource constraint equals the difference between the benefits of holding capital and the cost of holding capital.

Now let's try to write out the "expanded" version of the optimality condition for the control variable and solve for $c(t)$ in terms of the other variables. Note that $u' > 0$ for all c, which implies $\lambda(t) > 0$ for all t.

Starting with the optimality condition:

$$\lambda(t) = u'[c(t)]e^{-\rho t}$$

we seek to solve for $c(t)$ in terms of the other variables. We need to eliminate λ from the expression. Take the time derivative of each side to obtain:

$$\dot{\lambda}(t) = -\rho u'[c(t)]e^{-\rho t} + u''[c(t)]\dot{c}e^{-\rho t}.$$

Note that via the co-state equation constraint:

$$\dot{\lambda} = \lambda(t)[\delta - f'[k(t)]]$$

and via the optimality condition:

$$u'[c(t)]e^{-\rho t} = \lambda(t).$$

By making substitutions for $\dot{\lambda}$ and $\lambda(t)$, one obtains:

$$u'[c(t)]e^{-\rho t}[\delta - f'[k(t)]] = -\rho u'[c(t)]e^{-\rho t} + u''[c(t)]\dot{c}e^{-\rho t}.$$

Divide each side by $u'[c(t)]e^{-\rho t}$ to get:

$$[\delta - f'] = -\rho + \frac{u''}{u'}\dot{c}.$$

Rearranging gives us an expression in terms of dc/dt, which is:

$$\dot{c} = \frac{u'}{u''}[\delta + \rho - f'].$$

In other words, the optimal decision is to equate the rate at which consumption changes to a utility-weighted function of the rate of depreciation (population growth adjusted), the rate of impatience and the marginal productivity of capital. Note the obvious parallels with Eq. (5.8).

We might ask: what is the economic significance of $\lambda(t)$? In this case, $\lambda(t)$ is the shadow value of the capital constraint and the shadow value of a unit of capital. It tells us by how much the objective function changes in response to a unit change in the capital stock at time t. As such, it also tells us the tradeoff between consuming now and consuming later.

Additional Details

Recall that the Euler equation used in the calculus of variations was a single, second-order differential equation in the state variable. In contrast, the maximum principle uses two first-order differential equations: one in the state variable and one in the co-state variable. In addition, we require that the Hamiltonian be maximized with respect to the control variable. Actually, for problems with multiple state variables and multiple controls, we will have correspondingly more differential equations, but they will generally all be of the first order.

The first-order condition says that the Hamiltonian must be maximized with respect to the control path. In other words:

$$H(t, x, u^*, \lambda), \ \geq H(t, x, u, \lambda) \forall \, t \, in \, [0, T]. \tag{8.5}$$

This is the origin of the name "the Maximum Principle." As noted earlier, "the Maximum Principle" is a stronger statement than just $\partial H / \partial u = 0$. In fact, it may be the case that the condition $\partial H / \partial u = 0$ doesn't even apply. For example, if the control is bounded and the maximum occurs at a boundary of the control, then $\partial H / \partial u = 0$ won't hold. But for the specific case for which H is differentiable with respect to u and has an interior solution (the "standard" case, as noted), we require $\partial H / \partial u = 0$.

Note also that if either the terminal state or time is free, we will require a transversality condition. If the terminal time is fixed but the state is free, the corresponding transversality condition is:

$$\lambda(T) = 0. \tag{8.6}$$

This just says the shadow value of changes in the final stock (note the capital T corresponding to the end of the problem) must be zero. If this were not the case, it would be advantageous to adjust the final stock.

Dorfman's Interpretation

In a 1969 article in the *American Economic Review*, Robert Dorfman provided an economic interpretation of the maximum principle, including an economic interpretation of each of the conditions required for optimality.[4] We discussed some of these arguments in Chap. 3, when we interpreted the objective function in a calculus of variation problem as a profit function. Following the same interpretation, here is the gist of Dorfman's argument, which played a major role in introducing dynamic optimization into economics.

Think of the problem of maximizing a flow of profits over the interval [0,T]:

$$\Pi = \int_{0}^{T} \pi(t, K, u) dt$$

subject to:

---

[4] Generating economic intuition for the conditions required for optimality can be a challenge. Fellow economist Richard Woodward has developed a YouTube video to help. See https://www. youtube.com/watch?v=vuugpS313Tk.

$$\dot{K} = f(t, K, u)$$
$$K(0) = K_0$$
$$K(T) \text{ free.}$$

If we incorporate the equation of motion into the objective function and plug in all of the optimal paths and values, the following expression is obtained:

$$\Pi = \int_0^T \left[H(t, K^*, u, \lambda^*) + K^*(t)\dot{\lambda}^* \right] dt - \lambda^*(T)K^*(T) + \lambda^*(0)K_0.$$

Note that at the optimum, we require:

$$\frac{\partial \Pi^*}{\partial K_0} = \lambda^*(0).$$

In other words, $\lambda^*(0)$ measures the sensitivity of total profit to the initial level of capital. If we had one additional unit of capital available at the start of the problem, optimal profit could be lifted by the amount $\lambda^*(0)$. Thus, $\lambda^*(0)$ can be interpreted as the shadow value of a unit of capital at the beginning of the planning horizon.

We also have:

$$\frac{\partial \Pi^*}{\partial K^*(T)} = -\lambda^*(T).$$

This partial derivative gives the negative of the rate of change in profit with respect to the optimal *terminal* capital stock. In other words, if we wanted to hold one extra unit of capital at the end of the problem, we would have to sacrifice $\lambda^*(T)$ units of total profit. So, $\lambda^*(T)$ measures the shadow value of the terminal time capital stock. Using analogous reasoning, $\lambda^*(s)$ is the shadow value of a unit of capital received at any time $t = s$. This increment to capital provides no addition to profit between 0 and s, but does provide value in the interval s to T.

Control Restrictions

Optimal control problems do not require the control to be unconstrained or continuously differentiable. For example, restrictions may be placed on the control that require it to be a piecewise continuous function. Consider the previous example (à la Takayama (1985)) of filling a bathtub of fixed capacity with water from a faucet in a minimal amount of time (thus, T is free and $x(T)$ is fixed). Let the control variable $u(t)$ represent the amount of water that can be discharged from the faucet at each instant in time. If one defines the maximum amount of water than can be run from the faucet at any point in time as one and the minimum as zero then the control is restricted by $0 \leq u(t) \leq 1$. The rather trivial solution to the problem of filling the bathtub in a minimum amount of time is to choose $u(t) = 1$ for the interval $0 \leq t < T$ and $u(t) = 0$ for $t = T$. Such problems, in which the control variable takes on boundary values, with one or more switches between the start and end of the problem, are said to have "bang-bang" solutions, since the control "bangs" against its boundaries.

Problems in which the control has "jumps" at a finite number of points along the optimal path are not uncommon in economics. Such problems often arise when:

1. The Hamiltonian is linear in the control variable
2. The first-order condition $\partial H / \partial u = 0$ cannot be maintained at all times.

In economic applications, such conditions reflect situations in which policies must quickly and dramatically change. Somewhat surprisingly, there is at present, no general theory for dealing with such problems. Each case is distinct and methods of solution are often guided by intuition and experience, as the next example demonstrates.

Example: A "Bang-Bang" Optimal Control Problem

The following example is adapted from Kamien and Schwartz (1991). It is a good example of how optimal control provides a way to impose constraints, something which is not available in the calculus of variations. It also demonstrates that intuition and reasoning are sometimes needed in cases where calculus fails us.

We want to maximize:

$$V = \int\limits_{0}^{3} (2x - 3u)dt$$

subject to:

$$\dot{x} = x + u$$
$$x(0) = 4$$
$$x(3) \text{ free}$$
$$u(t) \in [0,2].$$

Two things are worth highlighting immediately. First, notice that the objective is linear in the control. We encountered a similar situation in Chap. 2, where Mr. Kuchenfresser's marginal utility of consumption was constant. In that case, we found non-uniqueness of the solution because the utility was linear in consumption. Here too, due to a lack of curvature, we can expect to face a similar difficulty in that we may not be able to find an interior solution to the problem. Since, the objective is linear in the control, one might expect that the control will spend part of its time at one or more bounds in the control space. Second, notice that we've imposed a constraint on the control variable, which in this case must be bounded between 0 and 2. Such imposition is not possible in the calculus of variations but can be accommodated in optimal control.

Let's proceed with what we know. If we incorporate the equation of motion into the objective function and construct the Hamiltonian, we obtain:

$$H = 2x - 3u + \lambda(x + u)$$

or after regrouping:

$$H = (2 + \lambda)x + (\lambda - 3)u.$$

Following known steps, and taking the derivative of the Hamiltonian with respect to the control and setting it equal to zero, we get:

$$\frac{\partial H}{\partial u} = \lambda - 3 = 0$$

which implies $\lambda = 3$.

As it turns out, this result is highly misleading and arises because the Hamiltonian is linear in the control. Recall that our condition for optimality is that u maximizes H. However, $\partial H / \partial u = 0$ is not required. We can reason as follows.

If $\lambda > 3$, then $\lambda - 3 > 0$ and $H = (2 + \lambda)x + (some\ positive\ value) \times u$.
If $\lambda < 3$, then $\lambda - 3 < 0$ and $H = (2 + \lambda)x + (some\ negative\ value) \times u$.

In the first case, the best way to maximize H is to set u as large as possible. In the second case, the best way to maximize H is to set u as small as possible. Because u is constrained to lie between 0 and 2, the "optimal" control in this case will be to set u to its maximum of 2 and then to its minimum of 0. In other words:

$$u^*(t) = \begin{cases} 2 & if\ \lambda(t) > 3 \\ 0 & if\ \lambda(t) \leq 3. \end{cases}$$

Now let's solve for

$$\dot{\lambda} = \frac{-\partial H}{\partial x} = -2 - \lambda$$

or

$$\dot{\lambda} + \lambda = -2.$$

This is a non-homogeneous first-order differential equation of the form $\frac{dy}{dt} + ay = b$, which has the solution form $Ae^{-at} + \frac{b}{a}$. So, with $a = 1$ and $b = -2$, we have:

$$\lambda(t) = ke^{-t} - 2.$$

We can use the endpoint restriction $\lambda(T) = \lambda(3) = 0$ to solve for:

$$0 = ke^{-3} - 2$$

to obtain:

$$k = 2e^3$$

and

$$\lambda(t) = 2e^{3-t} - 2.$$

This implies $\lambda(t)$ is a decreasing function of t.

Solving for the starting and ending values of λ, we get:

$$\lambda(0) = 2e^{3-0} - 2 = 38.171$$

and

$$\lambda(3) = 2e^{3-3} - 2 = 0.$$

At the start $\lambda > 3$ and at the end $\lambda < 3$. When $\lambda = 3$, the control switches (between 2 and 0, as we have seen). We can find the exact moment of the switch, call it $\hat{t}$, by solving for $\lambda = 3$:

$$\lambda(\hat{t}) = 3 = 2e^{3-\hat{t}} - 2$$

which implies

$$\hat{t} = 2.083.$$

The time paths of λ^* and u^* are displayed in Figs. 8.1 and 8.2.

Now, what can we surmise about x^* in the two regions delineated by $\hat{t} = 2.083$ and $\lambda = 3$? A bit more work is needed to fill out the picture completely.

We first note that $x(0) = 4$ is given and the co-state equation is given by $\dot{x} = x + u$. In the region in which $u^* = 2$, therefore, $\dot{x} = x + 2$; and in the region

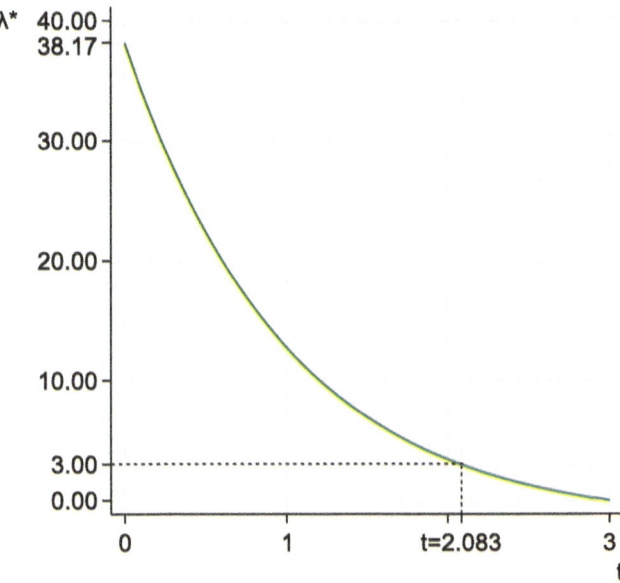

Fig. 8.1 Time path for λ^* in the "bang-bang" control problem

in which $u^* = 0$, $\dot{x} = x + 0$. Following the same procedure used above for solving differential equations of the form $\frac{dy}{dt} + ay = b$, and using the endpoint $x(0) = 4$ to define the initial constant, we can solve to obtain:

$$x(t) = 6e^t - 2.$$

This has the trajectory shown in Fig. 8.3.

Although this example hasn't been given any direct economic interpretation, keep in mind that bang-bang control problems do occur in the real world, for example, in situations in which control variables are constrained between a maximum and a minimum or when switching costs make using intermediate levels of controls suboptimal. Often, a firm's investment decision is lumpy ("all or nothing"). As another example, history shows that during an economic crisis, a central bank may rapidly move the government's interest rate down to, or close to, zero (the lower bound).[5] Then, once the economy recovers and begins to heat up, the rate is increased.

---

[5] Zero might seem like a reasonable lower bound on interest rates, but both the Bank of Japan and the European Central Bank have experimented with *negative* interest rates during periods of severe economic downturn. The idea is that by charging commercial banks a fee to store their money with the central bank, rather than paying them interest, commercial banks will choose to lend more to consumers and businesses, thereby stimulating economic activity.

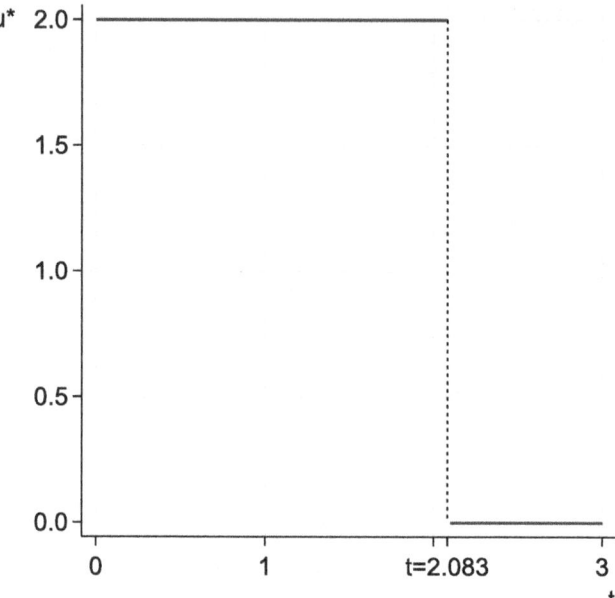

Fig. 8.2 Time path for u^* in the "bang-bang" control problem

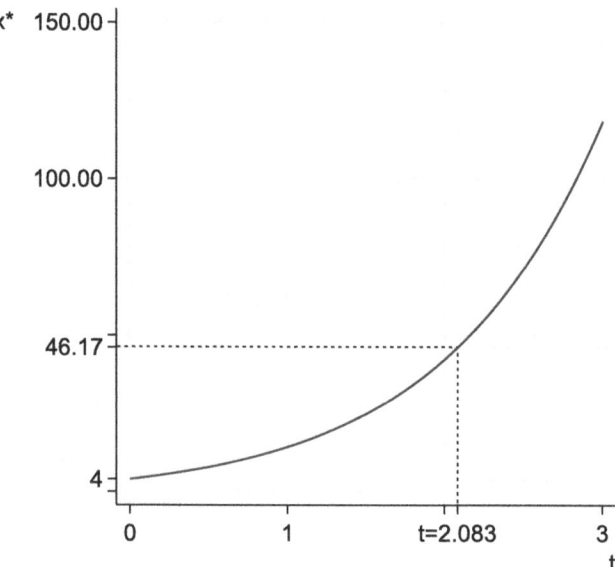

Fig. 8.3 Time path for x^* in the "bang-bang" control problem

References

Dorfman, R. (1969). An economic interpretation of optimal control theory. *American Economic Review, 59*, 817–831.

Kamien, M. I., & Schwartz, N. L. (1991). *Dynamic optimization: The calculus of variations and optimal control in economics and management* (2nd ed.). North-Holland Press.

Leonard, D., & Van Long, N. (1992). *Optimal control theory and static optimization in economics*. Cambridge University Press.

Mangasarian, O. L. (1966). Sufficient conditions for the optimal control of nonlinear systems. *SIAM Journal of Control, 4*, 139–152.

Takayama, A. (1985). *Mathematical economics* (2nd ed.). Cambridge University Press.

Vinter, R. (2010). *Optimal control. Modern Birkhäuser Classics*. Birkhäuser and Springer Nature.

Further Refinements in Optimal Control

9

As you explore increasingly sophisticated dynamic economic problems, you will likely encounter a wide range of examples developed using optimal control. This includes special cases and applications that deviate from the standard setup presented thus far. This chapter briefly reviews a few special cases and applications.

Transversality Conditions Revisited

As noted briefly at the end of Chap. 5, transversality conditions apply to problems when endpoints are not exactly specified and some flexibility exists with respect to how a problem starts or ends. When endpoint conditions are not specified, they become matters of choice, which introduces new variables into the problem. These new variables can apply to the initial or terminal times, the initial or terminal values of state variables, or the presence of a salvage value.

In addition, a combination of other kinds of restrictions can apply, including a range of inequality constraints. This leads to a rather large catalog of possible situations and transversality conditions. Leonard and van Long (1992), for example, provide a general formula for transversality conditions and outline 12 common cases in mathematical detail. Even their list is not exhaustive, however. In general, a transversality condition must be satisfied along with the standard optimality conditions associated with the calculus of variations or optimal control. They do not replace the standard first-order conditions, nor are they always necessarily binding. But if the starting points (stage and states) and ending points (stage and states) are not precisely defined in the problem, then these conditions, which replace the endpoint restrictions, must be satisfied. The new choice variables, of which there could be many, must be chosen optimally.

In most economic applications, the initial time and state are fixed, as illustrated in Fig. 9.1, where the path is indeterminant but required to move from a defined starting point (x_0, corresponding to starting time and initial state) to a specified ending point (x_T, corresponding to ending time and final state). For example, a firm may wish to completely exhaust its inventory of a specific product within a fixed period of time. The method and rate of doing so must be chosen to meet this schedule. Similarly, a traveling salesman may be required to be in a certain city on a certain date, but his route and speed of travel may be a matter of choice.

More generally, three very basic cases in which endpoints are flexible, and therefore, transversality conditions are needed, can be classified as in Table 9.1. Each is briefly highlighted below in the context of the Hamiltonian function of optimal control.

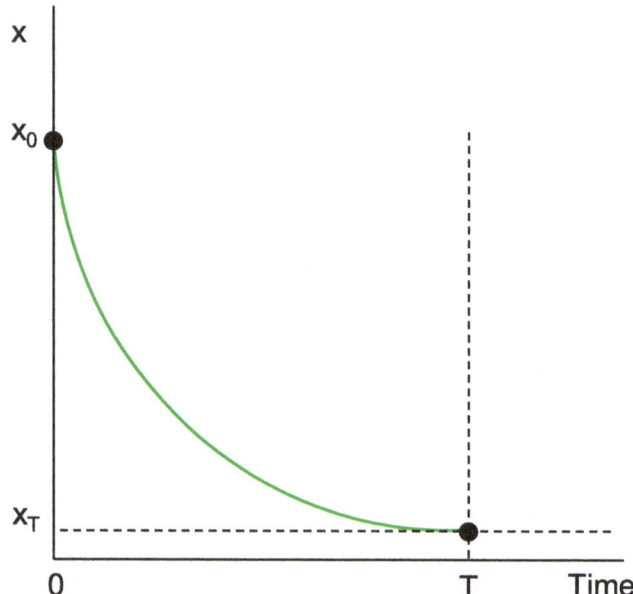

Fig. 9.1 The time path in a fixed endpoint problem

Table 9.1 Common situations requiring transversality conditions

Case	Terminal time	Terminal state
1	Free	Fixed
2	Fixed	Free
3	Free	Free

Terminal Time Free and Terminal State Fixed (Case 1)

This scenario is illustrated in Fig. 9.2. With a free terminal time, an extra condition is required to ensure the chosen ending time of the problem cannot be improved upon:

$$H(x^*(T^*), u^*(T^*), \lambda^*(T^*), T^*) = 0 \tag{9.1}$$

The interpretation of this condition is fairly straightforward. When the terminal time is free and can be extended by a small amount (say ΔT), the value of the integral of the fixed-time problem increases (if the Hamiltonian at that point is positive) or decreases (if the Hamiltonian is negative). The amount of the change is given by the amount by which the value function changes, i.e., $\partial V / \partial T = H$, and it follows that this derivative, evaluated at T^*, must be zero. For example, imagine that a firm is operating a mine on public land under a lease from the government. If the value function measures the net present value of the firm's profits, then when T is fixed, the firm's goal will be to squeeze as much value out of the mine as possible before the lease expires, and it will plan operations so that they remain profitable up to that point. If the firm is able to extend the lease, and thereby the period of operations, then there may be additional profit to be gained. The extension of operations may result in additional revenue but will also lead to additional cost. If the mine operator is free to choose how long to operate, it

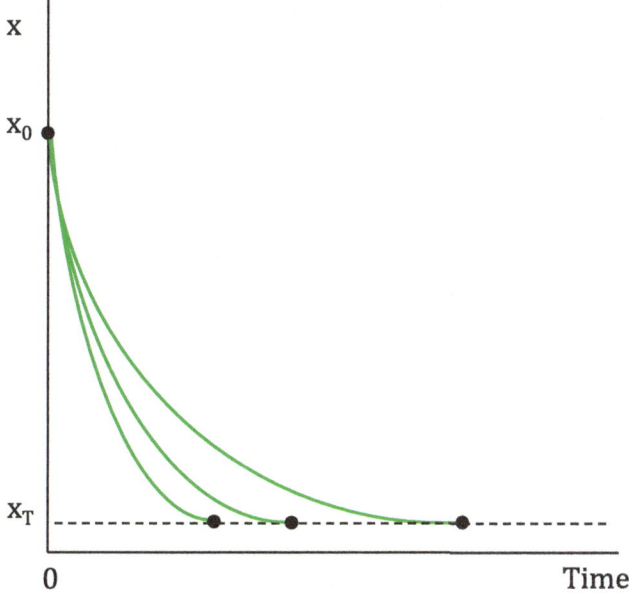

Fig. 9.2 Potential paths with a free terminal time and fixed terminal state

should be clear that it makes sense to operate as long as the mine is profitable, but not beyond the point where it is optimal. In other words, the profit added in the last period must be vanishingly small, such that, $\partial V/\partial T = 0$. Note that, if $\partial V/\partial T > 0 \ \forall \ T$, the optimal horizon is infinite: if the mine can remain profitable forever, then the firm should never shut it down.

Terminal Time Fixed but Terminal State Free (Case 2)

This scenario is illustrated in Fig. 9.3. With a free terminal state, the additional required condition ensures that any remaining amount of the state variable at the end of the problem cannot contribute to improvements in the value function:

$$\lambda(T^*) = 0. \tag{9.2}$$

Recall that $\lambda(t)$ is the costate variable associated with the state variable $x(t)$, i.e., the shadow value of the state variable. The interpretation of this transversality condition is therefore again fairly straightforward. If the state variable can be adjusted such that more or less of it remains at the terminal time, then it's shadow value (in terms of contribution to the value function) at that point should be zero. Imagining for a moment that the state variable again corresponds to an

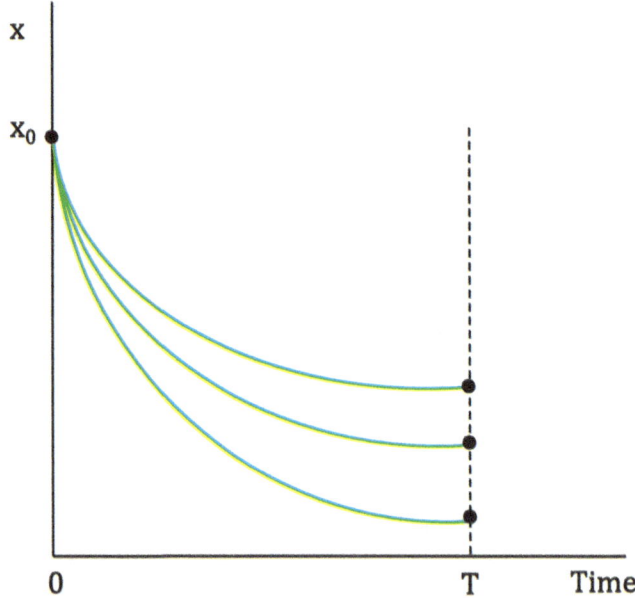

Fig. 9.3 Potential paths with a fixed terminal time and free terminal state

exhaustible asset, say ore in a mine, the firm should use the stock until its marginal contribution at the end of the planning horizon is zero. Note that this does not mean the stock should be driven to zero, since the cost of extracting the last unit may exceed its contribution to profit. Thinking back to Mr. Kuchenfresser, his friends and his cake, it should only be the case that cake remains at the end of the party if everyone is too full for another bite. Otherwise, it makes sense to continue eating. Mathematically, this transversality condition ensures that the optimal control problem satisfies the boundary conditions at the final time while maximizing or minimizing the objective function over the entire time interval.

Terminal Time and Terminal State Both Free (Case 3)

This scenario is illustrated in Fig. 9.4. With both a free terminal time and a free terminal state, two extra conditions are needed, namely, both of those specified above:

$$H(x(T^*), u(T^*), \lambda(T^*), T^*) = 0 \text{ and } \lambda(T^*) = 0 \tag{9.3}$$

When the terminal time is free and the terminal state is free, it should be the case that the value function cannot be improved further by adjusting either the ending time or state. In some problems, the final time and state may need to match a

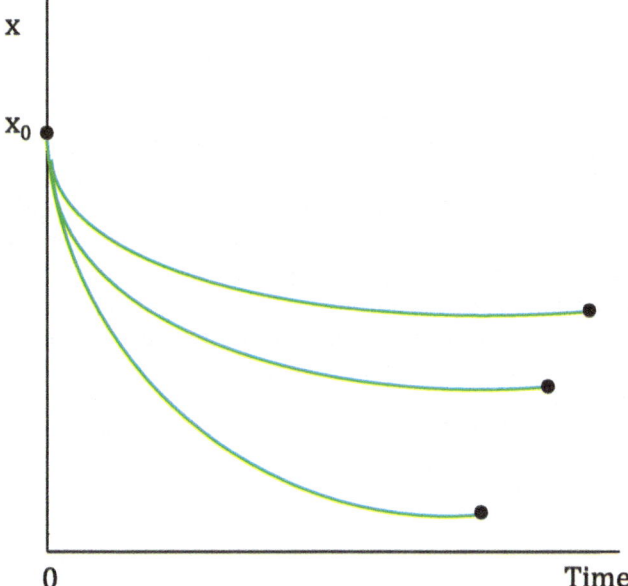

Fig. 9.4 Potential paths with a free terminal time and free terminal state

terminal line or **terminal surface**. In this kind of problem, which is a variation on Case 3 above, neither the final state nor the final time is prespecified, but neither are they completely free. For example, in seeking to finish one's Ph.D. dissertation, there may be a tradeoff between quality and time to completion. Faster completion may require that one sacrifice some quality. Conversely, aiming for high quality may require a longer time to complete one's work. This kind of problem can be handled by introducing an implicit function that ties the terminal time and terminal state together.

One of the most common complications to a dynamic optimization problem, beyond those outlined above, is the introduction of a salvage value that applies to the state variable at the end of the problem.[1] For example, it may be the case that a productive asset can be sold for scrap at the end of the problem. Understandably, the freedom of choosing when the problem ends combined with the possibility that the remaining value of the state variable contributes in a scalar fashion to the value function, means that the chosen terminal time must be optimal in both its impact on the Hamiltonian and in its impact on the value of the salvage function, which may itself decline as time goes by. The transversality condition in the case of a free terminal time with a salvage value is:

$$H^*(x^*(T^*), u^*(T^*), \lambda^*(T^*), T^*) + \frac{\partial S}{\partial T^*} = 0 \qquad (9.4)$$

where S represents the salvage value.

Control Parameters

Transversality conditions specify the characteristics of the dynamic system at the starting and stopping points. Importantly, transversality conditions should not be confused with constraints on the control variable. In Chap. 8, a constraint was placed on the control variable, which bound it between 0 and 2. That chapter ended with some brief examples in which, in some economic problems, restrictions may be imposed on the actions of a decision-maker. Such a constraint on the control variable is usually referred to as a **control parameter**. For example, consider a firm that has decided to build a new operating facility and must choose the size of the plant. Although the size of the plant may be a choice variable, it may not be infinitely variable. Furthermore, once the factory is built, plant capacity must remain fixed over the planning horizon. This problem may be thought of as an optimal control problem with capacity as a control parameter.

---

[1] Constraints at the start of the problem may also apply, although these are relatively rare in economics. For example, an initial cost incurred at the start of the problem could be incorporated as a negative salvage value.

In some cases, it may be possible to reformulate a problem with a control parameter into a problem with an additional state variable. For example, in the case of the factory decision, plant capacity could be defined as the state variable $z(t)$, with a new equation of motion, say $\dot{z}(t) = 0$ with free initial and terminal values. Although much more could be said about control parameters, they are mostly beyond our current scope. However, before going on, it is important to note that the standard optimal control problem assumes continuity of the control variable, which is in itself a constraint on the control variable. Problems that are linear in the control variable often have discontinuities in the control. For example, as we saw in Fig. 8.2, the control variable in a bang-bang problem does not follow a smooth path but rather switches between a maximum and a minimum value. These discontinuities correspond to what would be considered corner solutions in static optimization problems.

Blocked Intervals

Recall the thought experiment that concluded Chap. 1. Depending on your method of transport, when commuting from home to campus or work, you may encounter problems with your standard route: a sidewalk may be closed due to construction, a bus may break down, or an accident may temporarily close an intersection. The term **blocked interval** as applied to economic problems originates with Arrow (1968), who introduced the term to describe a situation in which constraints on a control variable prevent the path of the state variable $x(t)$ from following the singular path $x^*(t)$. An example is the case in which the state variable is associated with a non-negativity

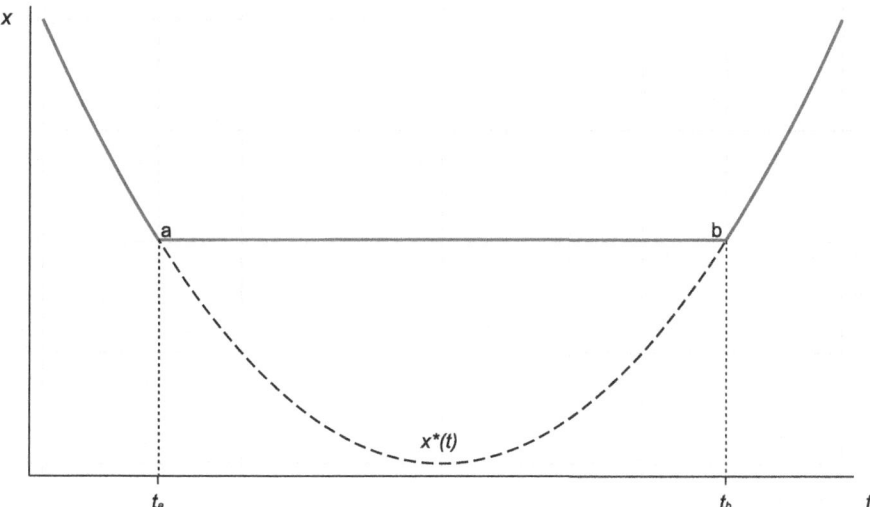

Fig. 9.5 A blocked interval between t_a and t_b for the path $x^*(t)$

constraint; say $\dot{x} \geq 0$. If x is defined as the capital stock and investment is constrained to be irreversible, it might be the case that the firm can buy capital but cannot sell installed capital. Figure 9.5 provides a simple illustration of a blocked interval, where the curved dotted line is the optimal path $x^*(t)$ and the blocked interval occurs between t_a and t_b. Because the path is blocked during the interval (t_a, t_b), instead of following its preferred path $x^*(t)$, the path follows the line segment $\overline{ab}$ in the blocked interval, and therefore the green solid line for its entire length.

Blocked interval problems are usually solved by breaking down the problem into its constituent pieces, optimizing the value function with respect to the time at which the blocked interval begins, and then again with respect to the time beginning when the blocked interval ends. Solutions for the control variable typically involve what is known as **premature switching**. For some further discussion and examples, see Clark (1990).

Current-value Hamiltonians

Discounting appeared in Chap. 3, where impatience was incorporated into Mr. Kuchenfresser's problem. While it would appear that economists are on fairly solid philosophical ground when including a discount factor in intertemporal problems, in fact discounting did not appear explicitly in early economic writings. For example, Alfred Marshall relegates the concept of discounting to hedonics, "not properly to economics." Similarly, in his (1928) paper on optimal intertemporal savings rates, Frank Ramsey rejected the use of discounting on ethical grounds. Until the 1960s, almost without exception, this was the rule: discounting was not a central feature in intertemporal problems. A primary exception is Hotelling (1931), who examined the optimal extraction of a mineral reserve over time.

In the 1960s, as economics became more formally associated with mathematics, economists began investigating the relationship between basic postulates regarding utility functions and consumer preferences over time. In this work, they showed that a utility function exhibiting time neutrality could not represent all possible consumption paths, and therefore that time preference for the present was a mathematical necessity for solving many intertemporal economic problems.[2]

Although discounting has already been encountered in previous problems, here the issue is treated more formally. To incorporate discounting into an optimal control problem, the mathematical framework must be developed a bit further. Consider the general problem:

---

[2] Philosophers (and some economists) have pointed out that discounting tends to blur the line between efficiency concerns and equity concerns. Ethical debates regarding the appropriateness of discounting in economics, especially for social concerns, continue. Some of these arguments are summarized by Parfit (1984).

$$Max \int_0^T e^{-rt}F(x, u, t)dt$$

subject to:

$$\dot{x} = G(x, u, t)$$

$$x(0) = 0$$

where x is the state variable, u is the control variable, F is a value function, and G is a constraint equation. Values are discounted at the rate r.

The Hamiltonian for this problem is:

$$H = e^{-rt}F(x, u, t) + \lambda G(x, u, t)$$

and the requirements for an optimum are:

$$\frac{\partial H}{\partial u} = e^{-rt}\frac{\partial F}{\partial u} + \lambda\frac{\partial G}{\partial u} = 0$$

$$-\frac{\partial H}{\partial x} = \dot{\lambda} = e^{-rt}\frac{\partial F}{\partial x} - \lambda\frac{\partial G}{\partial x}$$

$$\frac{\partial H}{\partial \lambda} = G(x, u, t)$$

$$\lambda(T) = 0.$$

Values that appear here are discounted back to time 0. In particular, the multiplier $\lambda(t)$ gives a marginal value for the state variable at time t, discounted back to time 0.

It sometimes proves more convenient to express the Hamiltonian in a current-value form:

$$H = e^{-rt}[F(x, u, t) + e^{rt}G(x, u, t)] \qquad (9.5)$$

and to define a new, **current-value multiplier**:

$$m(t) \equiv e^{rt}(t). \qquad (9.6)$$

These can be interpreted as follows: $\lambda(t)$ is the marginal value of the state variable at time t, discounted back to time 0; in contrast, $m(t)$ provides the marginal value of the state variable at time t in terms of value at time t.

Using (Eq. 9.6), the **current-value Hamiltonian** can be rewritten as:

$$\widehat{H} = F(x, u, t) + mG(x, u, t). \tag{9.7}$$

To see the correspondence between the present value and current value Hamiltonians, differentiate (Eq. 9.6) with respect to time to get:

$$\begin{aligned}\dot{m} &= re^{rt}\lambda + \dot{\lambda}e^{rt} \\ &= rm - e^{rt}\tfrac{\partial H}{\partial x}\end{aligned} \tag{9.8a}$$

The definition implies $H = \hat{H}e^{-rt}$, so equation (Eq. 9.8a) is:

$$\dot{m} = rm - \frac{\partial F}{\partial x} - m\frac{\partial G}{\partial x}. \tag{9.8b}$$

The present value first-order condition can be rewritten:

$$\frac{\partial H}{\partial u} = e^{-rt}\frac{\partial \hat{H}}{\partial u} = 0. \tag{9.9}$$

Eq. (9.9) implies the optimality condition: $\partial \hat{H}/\partial u = 0$. Furthermore, the equation of motion can be recovered in terms of the current value Hamiltonian as:

$$\dot{x} = \partial \hat{H}/\partial m = G. \tag{9.10}$$

The present value Hamiltonian and the necessary conditions for optimality can therefore be re-stated in current value form as:

$$\hat{H} = F(x, u, t) + mG(x, u, t) \tag{9.11}$$

$$\frac{\partial \hat{H}}{\partial u} = \frac{\partial F}{\partial u} + m\frac{\partial G}{\partial u} = 0 \tag{9.12}$$

$$\dot{m} = rm - \frac{\partial F}{\partial x} - m\frac{\partial G}{\partial x}. \tag{9.13}$$

Transversality conditions also may be stated in current value form. For example, $x(T)$ free requires $e^{-rt}m(T) = 0$ and $x(T) = 0$ requires $e^{-rt}m(T) = 0$ as well as $e^{-rt}m(T)x(T) = 0$.

One might ask: why should we bother with the current value setup? As it turns out, using the current value Hamiltonian simplifies finding solutions in many settings. The first advantage is that the current value first-order conditions do not contain any discounted terms. The second and main advantage is that if F and G do not contain t as explicit arguments, then the current value first-order conditions reduce to a set of autonomous equations, that is, they do not depend on time explicitly. Autonomous differential equations are much easier to solve than non-autonomous equations and can be more easily plotted since their paths can be plotted in (x, u) space independent of time.[3]

Example: Capital Investment

Eisner and Strotz (1963) introduced a basic model of capital investment with adjustment cost. Their setup assumes that the adjustment cost function is strictly convex and quadratic (that is, the cost per unit of investment rises with the investment rate). Although the original version of the Eisner-Strotz model is written in the calculus of variations, it is easily adapted to an optimal control framework. The problem is to:

$$Max \int_{t=0}^{t=T} [pQ(L(t), K(t)) - wL(t) - C(I(t))]e^{-rt}dt$$

subject to:

$$\dot{K} = I(t) - \delta K(t)$$
$$L(t) \geq 0$$
$$K(t) \geq 0$$
$$K(0) \geq K_0$$

where $\dot{K}$ represents the rate of investment, $Q(L, K)$ is a production function that uses labor (L) and capital (K), w is the wage rate for labor, and $C(I)$ is the cost of investment function. Assume an upward sloping cost function, i.e., that $C > 0$, $C' > 0$, and $C'' > 0$. The last condition means adjustment costs will be increasingly greater the greater the rate of investment. This form of curvature ensures some friction in the investment decision. Our intuition should tell us that the problem is likely to have an interior solution where the marginal benefit of investment is balanced by the marginal cost of investment. The goal here is not necessarily to solve the problem, but rather to show how introducing discounting alters the setup.

The present value Hamiltonian for the problem is:

$$H(K, I, \lambda) = pQ(L(t), K(t)) - wL(t) - C(I(t)) + \lambda(t)[I(t) - \delta K(t)]$$

---

[3] The practical implication is that in phase diagrams the $dx/dt = 0$ locus does not shift over time. In a non-autonomous system, the $dx/dt = 0$ locus shifts over time, making phase diagram analysis virtually impossible.

where $\lambda(t)$ is the costate variable, i.e., the shadow value of capital at time t. This is called the "present value Hamiltonian" because the values of all variables are measured at the start of the problem. The optimality conditions for the present value formulation are:

$$\frac{\partial H(K,\ I,\ \lambda)}{\partial I} = -C'(I(t)) + \lambda = 0$$

and

$$\dot{\lambda} = r\lambda - \frac{\partial H(K,\ I,\ \lambda)}{\partial K} = (r+\delta)\lambda - p\frac{\partial Q(L(t), K(t))}{\partial K}.$$

Furthermore, since initial capital is given, but the terminal level is not, a transversality condition (such as that for Case 2 in Table 9.1) is needed:

$$\lambda(T) = 0.$$

Now, let's examine the current value version of the same problem. Remember, the key difference is that while the present value Hamiltonian uses the costate variable $\lambda(t)$ directly, the current value Hamiltonian replaces $\lambda(t)$ with a modified shadow value, namely:

$$m(t) = e^{rt}\lambda(t).$$

This modified shadow value, $m(t)$, takes the present value shadow value and compounds it forward in time at the rate r, to time t along the path. It tells us the shadow value of capital not at the start of the problem, but at the moment along the path when that unit of capital is realized.

The current value Hamiltonian for the problem is:

$$H(K,\ I,\ \lambda) = pQ(L(t), K(t)) - wL(t) - C(I(t)) + m(t)[\ I(t) - \delta K(t)]$$

where $\lambda(t)$ is simply replaced with $m(t)$. The new optimality conditions are:

$$\frac{\partial H(K,\ I,\ \lambda)}{\partial I} = -C'(I(t)) + m = 0$$

$$\dot{m} = rm - \frac{\partial H(K,\ I,\ \lambda)}{\partial K} = (r+\delta)m - p\frac{\partial Q(L(t), K(t))}{\partial K}$$

and

$$m(T) = 0.$$

Table 9.2 A comparison of present value and current value formulations

Component	Present value form	Current value form
Objective	$Max \int\limits_{t=0}^{t=T} [\pi - C] e^{-rt} dt$	Same
State equation	$\dot{K} = I - \delta K$	Same
Costate variable	$\lambda(t)$	$m(t)$
Hamiltonian	$H = \pi - C + \lambda(I - \delta K)$	$H = \pi - C + m(I - \delta K)$
FOC	$C' = \lambda$	$C' = m$
Costate equation	$\dot{\lambda} = (r + \delta)\lambda - \frac{\partial \pi}{\partial K}$	$\dot{m} = (r + \delta)m - \frac{\partial \pi}{\partial K}$
Transversality	$\lambda(T) = 0$	$m(T) = 0$
Interpretation	$\lambda(t)$ is the discounted marginal value of capital	$m(t)$ is the current marginal value of capital

This current value formulation is often preferred in economic problems, not only because of the solution strategies indicated previously, but also because the current value costate variable $m(t)$ can be directly interpreted as the marginal value of capital in present units.[4] Table 9.2 provides a side-by-side comparison of the present value and current value formulations, where, for simplicity, $\pi(K)$ takes the place of $pQ(L(t), K(t)) - wL(t) - C(I(t))$ in the statement of the problem above.

References

Arrow, K. J. (1968). Applications of control theory to economic growth. In G. B. Dantzig & A. F. Veinott (Eds.), *Mathematics of the decision sciences part 2*. American Mathematical Society.

Clark, C. W. (1990). *Mathematical bioeconomics: The optimal management of renewable resources*. Wiley.

Eisner, R., & Strotz, R. H. (1963). Determinants of business investment. In D. B. Suits, et al. (Eds.), *Impacts of monetary policy*. Prentice Hall.

Hotelling, H. (1931). The economics of exhaustible resources. *Journal of Political Economy*, *39*, 137–175.

Leonard, D., & Van Long, N. (1992). *Optimal control theory and static optimization in economics*. Cambridge University Press.

Parfit, D. (1984). *Reasons and persons*. Clarendon Press.

Ramsey, F. (1928). A mathematical theory of savings. *Economic Journal*, *38*(152), 543–559.

Strotz, R. H. (1955). Myopia and inconsistency in dynamic utility maximization. *Review of Economic Studies*, *23*, 165–180.

---

[4] It is a somewhat technical point, but using a discount factor other than e^{-rt} (or the discrete time analogue) could lead to a situation in which the planner would wish to change the optimal plan over time. This problem, identified as dynamic inconsistency by Strotz (1955), is the reason for nearly universal use in economics today of exponential discounting.

Dynamic Stability

10

If you continue to read and explore the literature on dynamic economic models, especially in the context of macroeconomics or natural resource economics, you will undoubtedly encounter the topic of **dynamic stability** and the discussion of phase diagrams. Phase diagrams can be a bit perplexing when studied for the first time, but like many things, they make more sense with repeated exposure. If you have ever looked at a topographical map of a mountain or a weather map indicating wind direction and speed, then you are already somewhat familiar with what a phase-diagram conveys. This chapter focuses first on deriving a phase diagram for a two-equation system of differential equations. We then look at the general methods for assessing the stability of a dynamical system. We don't dive too deeply into this material. Plenty of advanced treatments are available, including some of the sources already mentioned.[1]

A Phase-Plane Example

Some dynamic optimization problems can be fully characterized by an analytical solution. However, many problems of interest in economics cannot. In these cases, we sometimes revert to using a **phase-plane diagram** to provide a graphical perspective on how variables move in relation to one another. A phase-plane diagram is a graph that shows the movements of a dynamic system. We already encountered simple phase diagrams in Chap. 6, first in the context of the cobweb model (recall Fig. 6.1) and then in the context of capital accumulation (recall Fig. 6.2). The primary reason for using phase diagrams is

---

[1] For those who truly want to dig deeper into this material, the treatment by Caputo (2005) is highly recommended, where the static primal-dual comparative statics framework of Silberberg (1978) is extended to a dynamic context.

to get a qualitative solution to a problem for which an analytical solution may not exist. Another reason for deriving phase diagrams, and probably the more useful one, is to assess the **stability** of dynamical systems or their tendencies toward or away from stable locations or **steady states**.

Let us return to our example of a renewable natural resource from Chap. 6. Let's place our fishery in the hands of a monopolist who harvests fish while facing linear inverse demand $p(q)$. This means we can write revenue straightforwardly as price × quantity, i.e., $R(q) = p(q)q$. To keep the problem as simple as possible, let's assume that the cost of fishing is zero and the total catch in any period is unconstrained. Admittedly, these assumptions—especially the first—are rather unrealistic, but our goal at this point is to illustrate the workings of a phase diagram, not to model a fishery in complete detail.

The growth rate of the fish population, i.e., dx/dt, is $\dot{x}$, and growth follows the logistic form encountered earlier:

$$g(x, q, t) = ax - bx^2 - q \tag{10.1}$$

where x is the stock of fish, q is the harvest, and $a > 0$ and $b > 0$ are biophysical parameters which are assumed to remain constant. This implies $\partial g / \partial x = a - 2bx$, which you might recognize as a constant coefficient FODE.

The firm chooses the harvest rate $q(t)$ in each period to maximize the discounted stream of revenues $R(t)$, subject to the available fish stock $x(t)$. Define the present-value Hamiltonian as:

$$H(x, q, \lambda, t) = R(q)e^{-rt} + \lambda g(x, q, t). \tag{10.2}$$

The optimality condition is:

$$\frac{\partial H}{\partial q} = R_q e^{-rt} + \lambda g_q = 0. \tag{10.3}$$

The co-state equation is:

$$-\dot{\lambda} = \lambda g_x \tag{10.4}$$

and the state equation is just a restatement of the growth equation:

$$\dot{x} = g(x, q, t) = ax - bx^2 - q. \tag{10.5}$$

These equations provide the elements required to construct the phase diagram.

Constructing the Phase Diagram

The goal when constructing a phase diagram is to plot two differential equations, which together describe the dynamic properties of the system. In our current example, one of these equations will describe the growth dynamics of the fish population. The other will describe the behavioral dynamics of the economic agent harvesting from the population. Together, these will provide insights into how the two parts—the natural resource and the economic decision maker—interact. In order to construct this phase diagram, it is necessary to derive a system of two differential equations. Currently, our problem has three variables: q, x, and t, corresponding to harvest, growth, and time. Recall from past chapters that a first line of attack when solving a Hamiltonian is to eliminate a variable. We do the same thing here and eliminate the co-state variable using (Eqs. 10.3 and 10.4). This is done in several steps.

First note that $\partial g / \partial q = -1$. This means that (Eq. 10.3) can be written as:

$$R_q e^{-rt} - \lambda = 0. \tag{10.6}$$

If equation (Eq. 10.6) is differentiated with respect to time, it leads to:

$$\dot{\lambda} = -rR_q e^{-rt} + R_{qq}\dot{q}e^{-rt} \tag{10.7}$$

or

$$\dot{\lambda} = R_q e^{-rt}\left[\frac{R_{qq}}{R_q}\dot{q} - r\right]. \tag{10.8}$$

Now (Eqs. 10.4 and 10.6) can be used to solve out $\dot{\lambda}$:

$$\lambda g_x = -\lambda\left[\frac{R_{qq}}{R_q}\dot{q} - r\right]. \tag{10.9}$$

Simplifying (Eq. 10.9) leads to:

$$\dot{q} = (r - g_x)\frac{R_{qq}}{R_q}. \tag{10.10}$$

This is the first differential equation to be plotted. Note that when the rate of growth equals the interest rate (i.e., when $g_x = r$), $\dot{q} = 0$. Hold that thought.

The second differential equation is easy to get, since it is supplied directly by the growth equation:

$$\dot{x} = ax - bx^2 - q. \tag{10.11}$$

Recall that back in Chap. 6, we noted the value of working with autonomous differential equations. Fortunately, (Eqs. 10.10 and 10.11) are autonomous, meaning time does not appear in either of them directly. This means that the functional forms for the differential equations do not change over time, and we can examine their interactions graphically, abstracting from the point in time when the interaction might be taking place, since the functional relationships hold at all times. Equations (10.10 and 10.11) can be used in several ways: (i) to find the optimal harvest rate given initial conditions; (ii) to determine a steady-state solution to the problem; and (iii) to assess the stability of this steady state.

A **steady-state solution** to the problem is one in which neither the state variable nor the control variable is changing over time. In other words, there is no incentive for the economic agent to make adjustments and no movement in the underlying natural resource itself that leads to adjustments. The steady-state solution is said to exist where $\dot{x} = \dot{q} = 0$. Steady states may hold some appeal, but they are not guaranteed and may be difficult to maintain.

A phase-plane diagram for this problem can be constructed by identifying the lines along which $\dot{x} = 0$ and $\dot{q} = 0$ and plotting them in x–q space, with the state variable x on the horizontal axis and the control variable q on the vertical axis. This looks like Fig. 10.1, where $\dot{q} = 0$ is a vertical line along which the rate of growth in the stock equals the interest rate. To the left of this vertical line, the stock is less than the steady state level and to the right it is above the steady state level. Why might this be? To begin, one can reason that the identity $(r - g_x)R_{qq}/R_q = 0$ can only be meaningfully satisfied if $r = g_x$. The locus of points where $\dot{q} = 0$ therefore describes points at which no change in harvest is desired: the productivity of the stock equals the interest rate. If this were not the case, say if the stock were more productive than proceeds in the bank, then it would make sense to adjust the harvest a tiny amount to take advantage of this difference.

In the case of the line $\dot{x} = 0$, this can be derived from the stock growth equation. This equation gives a locus of points along which $\dot{x} = 0$. It connects the points where

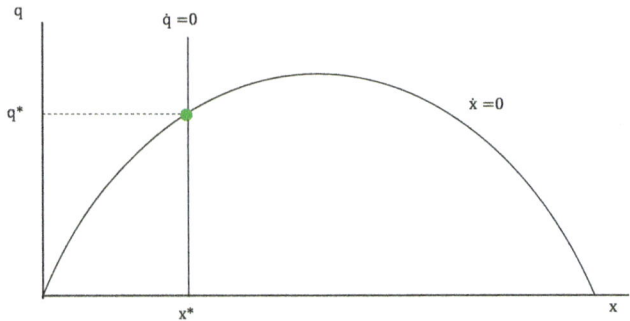

Fig. 10.1 The resource steady state where $\dot{x} = 0$ and $\dot{q} = 0$

harvest equals growth, and hence the growth rate is stationary. This condition implies that $ax - bx2 = q$ along the path. To find this curve, one can simply plot values of q for x. This appears in Fig. 10.1 as a parabola that intersects the x-axis twice, both times at the point where $q = 0$. The maximum of the function is attained where $x = a/2b$. This curve shows the points at which the stock is stable. Anywhere off this curve, the stock is not stable, and is either growing (more fish) or becoming smaller (fewer fish).

To reiterate, for intuition, you might think of the vertical line defined by $r = g_x$ as a no-arbitrage condition for the owner of the resource. If the interest rate were higher than the growth rate of the resource, then it would be advantageous to harvest more and invest the proceeds elsewhere. If the growth rate were higher than the interest rate, then it would be advantageous to harvest less and allow the resource stock to grow in place. Abstracting from the possibility that returns might be risky or uncertain, in a steady state, the resource owner matches the return on the resource to the return that can be obtained elsewhere. This gives some insight into why some natural resources with very slow growth rates have been exploited to or near extinction: if the outside return on investment provides a higher return than the resource itself, the resource owner may be incentivized to rapidly liquidate the resource. This is especially true when property rights over the resource are not well defined, and it is treated as an **open access resource**. Technically, however, slightly more stringent mathematical conditions are needed to ensure extinction of the resource, since the cost of harvesting the last few animals may be so prohibitive as to prevent total elimination of the stock.[2]

We are now in a position to begin thinking about the dynamics of the system defined by growth and extraction. First note that the lines in Fig. 10.1 are, technically, known as **isoclines**. An isocline (from the Greek *iso*, meaning "same," + *kline*, meaning "slope") is simply a curve on a graph along which all points have the same gradient. You can think of them as functioning like the contour lines on a topographical map, which indicate areas of equal steepness. An isocline in the phase diagram is a line where the differential equation has a constant value. The steady-state isoclines for this problem are the lines along which $\dot{x} = 0$ and $\dot{q} = 0$. In brief, the isoclines divide the xq space into separate regions. In each of these regions the stock and harvest trajectories move in specific directions. Recall that:

---

[2] A classic example of a resource hunted to extinction is the passenger pigeon (*Ectopistes migratorius*), which was once the most abundant bird in North America. In the nineteenth century, hunting pressure and habitat loss led to a collapse of the species. Despite last-ditch efforts to protect nesting populations, the final remaining wild bird was shot in southern Ohio in 1900. The last captive bird, Martha, died at the Cincinnati Zoo on September 1, 1914. In case you are interested, the sad and cautionary tale of the bird is told in *The Passenger Pigeon: Its Natural History and Extinction* (Schorger 1955). Remarkably, efforts are currently underway to use the genetic material of museum specimens to restore a population of similar, if not genetically identical, birds through the process known as de-extinction.

$$\dot{x} = 0 \ \text{ implies that } ax - bx^2 = q$$

and

$$\dot{q} = 0 \ \text{ implies that } (r - g_x)R_{qq}/R_q = 0.$$

The steady state for the system occurs where these two isoclines intersect. This is the point defined as the intersection (x^*, q^*) in Fig. 10.1.

For diagnostic purposes, think of the isoclines as dividing the graph into four separate regions, where the movement of the underlying variables is guided by different forces. Figure 10.2 reproduces Fig. 10.1 with these regions labeled I, II, III, IV, and with arrows added to indicate the expected movement off the isoclines. At points off of the isoclines, the **direction** of the system dynamics depends on the signs of the derivatives of $\dot{x}$ and $\dot{q}$. The **speed** of these adjustments depends on the magnitudes of the derivatives. Using the analogy of a topographical map, the sign of the derivative indicates whether moving off the current contour leads up or down hill. The size of the derivative indicates how steeply the ascent or descent might be. Moving off a contour where the derivative is large and negative results in a rapid fall, in the same way that stepping the wrong way off a steep trail when hiking in the mountains leads to a precipitous fall.

Interpreting the Phase Diagram

In Fig. 10.2, the isoclines divide the graph into four regions or sectors. Each is cross-hatched for ease of interpretation. Within each sector, the direction of movement of the system is the same. The dynamics of this particular system represent a **saddle point** centered at (x^*, q^*). It is called a saddle point because, in three dimensions, the diagram looks like a horse's saddle, bending upward along the horse's back from (x^*, q^*) in two directions (regions II and IV in Fig. 10.2) and bending downward across the horse's sides from (x^*, q^*) in opposite directions (regions I and III in Fig. 10.2).

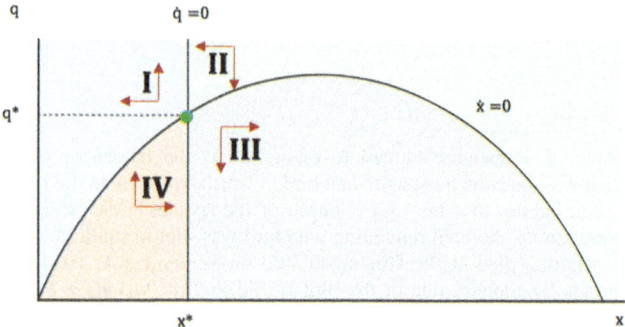

Fig. 10.2 Phase diagram with isoclines and directions of movement

In other words, if displaced from the steady state, some points lie on a path (a **trajectory**) leading back to the steady state and some points lie on a path leading away from the steady state. For example, in Region I, harvest exceeds the steady state harvest, i.e., $q > q^*$, and the stock is below the steady state stock, i.e., $x < x^*$. This is clearly an unsustainable situation that will lead to a collapse in the stock. Why might such a situation arise? As reasoned earlier, if the internal growth rate of the resource is less than the interest rate, the resource manager prefers to exploit the resource, since the rate of return from moving the capital to an alternative use is greater than the inherent growth rate of the resource. In Region IV, in contrast, while the stock is also below the steady state stock, i.e., $x < x^*$, harvest is below the steady state harvest, i.e., $q < q^*$, providing an opportunity for the stock to grow toward the steady state level. Movements in other regions can be similarly reasoned. This example is illustrative but rather simplistic. Neher (1990) provides a more rigorous treatment of the complexities surrounding optimally managing a fishery, which can prove exceedingly difficult.

Perturbations From the Steady State

As we have seen, if a system is at its steady state but receives some kind of "shock," one might naturally want to know what happens when the shock displaces it from the steady state. This question is important because it tells us something about whether the system might converge to a steady state or diverge away from it. We saw this argument in the context of the cobweb model example in Chap. 6, where a shock that moved supply and demand out of balance led to an iterative return to equilibrium (although such a result was not guaranteed). It is also important to know about system stability, because knowing whether a system is stable or unstable may help us to formulate policy: an unstable system might need much more careful management or protection than a stable system. Phase diagrams are one approach to looking into this question.

A more formal approach to addressing this question takes us into some new mathematical territory. Here we follow Kamien and Schwartz (1991) in sketching out the basic technique, which turns out to be somewhat simpler and more elegant than reasoning one's way through a phase diagram.

Begin by writing the differential equations for the system as functions of their arguments, namely $\dot{q} = v(x, q)$ and $\dot{x} = w(x, q)$. We have already established that the steady state occurs where $\dot{q} = 0$ and $\dot{x} = 0$, i.e., where $v(x^*, q^*) = 0$ and $w(x^*, q^*) = 0$. If v and w are relatively smooth in the neighborhood of $\dot{q} = 0$ and $\dot{x} = 0$, they can be approximated by first-order Taylor series expansions:[3]

---

[3] For those encountering Taylor series expansions for the first time, a bit of parallel study from a reputable source might be warranted. In brief, a Taylor series provides a way to approximate a non-linear function with a linear function, making it easier to work with. They've been around since the English mathematician and Royal Society member Brook Taylor proposed them in 1715. Evidence of Taylor's impact can be found on the moon, where a crater is named in his honor.

$$\dot{q} \cong v_x(x - x*) + v_q(q - q*)$$

and

$$\dot{x} \cong w_x(x - x*) + w_q(q - q*).$$

The partial derivatives can be collected into a two-by-two matrix. Let's call this matrix of partial derivatives $\mathbf{B}$:

$$\mathbf{B} = \begin{bmatrix} v_x & v_q \\ w_x & w_q \end{bmatrix}.$$

The stability of a dynamic system can be determined on the basis of an **eigenvalue** test. This test relies on the numerical values of the eigenvalues of $\mathbf{B}$, the matrix of partial derivatives of the differential equations, which must be evaluated in the neighborhood of the steady state.[4] A wide range of stability (and instability) conditions can be identified in this way. Examples of some of the most frequently encountered stability conditions are summarized in Table 10.1 and graphed in Fig. 10.3. Guckenheimer and Holmes (1983) cover this material in fine detail.

Table 10.1 Stability conditions and Eigenvalue tests

Figure 10.3	Type of equilibrium	Eigenvalue test	Kamien and Schwartz (1991)
1	Linear trajectories		
1a	Stable asymptotic trajectory	$\phi_1 < 0, \phi_2 < 0$	Case IA
1b	Unstable asymptotic trajectory	$\phi_1 > 0, \phi_2 > 0$	Case IB
2	Non-linear trajectories		
2a	Stable asymptotic trajectory	$\phi_1 < 0, \phi_2 < 0$	Case ID
2b	Unstable asymptotic trajectory	$\phi_1 > 0, \phi_2 > 0$	Case IE
3	Conditionally stable saddle point	$\phi_1 > 0, \phi_2 < 0$ $\phi_1 < 0, \phi_2 > 0$	Case IC
4	Spiral	$\phi_1, \phi_2 \in I$	
4a	Asymptotically stable	$\text{Real}(\phi_1) < 0, \text{Real}(\phi_2) < 0$	Case IIB
4b	Unstable	$\text{Real}(\phi_1) > 0, \text{Real}(\phi_2) > 0$	Case IIC

---

[4] Eigenvalues (denoted ϕ in Table 10.1) are also called characteristic roots, latent roots, or λ-roots. They are the roots of the characteristic equation of a square matrix. These roots may be real or imaginary (denoted I in Table 10.1). Note that if the eigenvalues of a dynamic system are zero, the system's stability is uncertain.

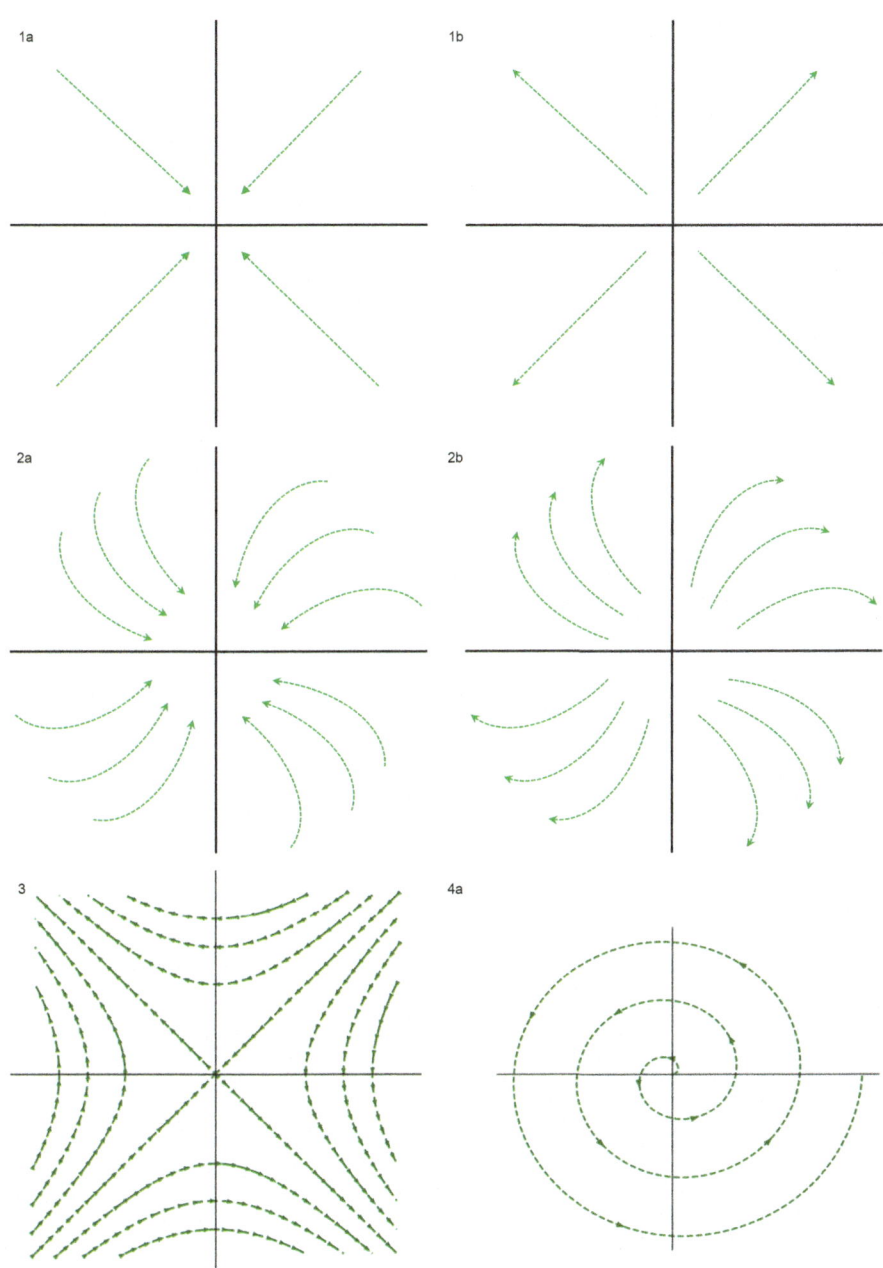

Fig. 10.3 Stability diagrams; see text for description of trajectories

References

Caputo, M. R. (2005). *Foundations of dynamic economic analysis*. Cambridge University Press.

Guckenheimer, J., & Holmes, P. (1983). *Nonlinear oscillations, dynamical systems, and bifurcations of vector fields (applied mathematical sciences* (Vol. 42). Springer-Verlag.

Kamien, M. I., & Schwartz, N. L. (1991). *Dynamic optimization: The calculus of variations and optimal control in economics and management* (2nd ed.). North-Holland Press.

Neher, P. A. (1990). *Natural resource economics: Conservation and exploitation*. Cambridge University Press.

Schorger, A. W. (1955). *The passenger pigeon: Its natural history and extinction*. University of Wisconsin Press.

Silberberg, E. (1978). *The structure of economics: A mathematical analysis*. McGraw-Hill.

Dynamic Programming

11

The weather is clear and bright, and Mr. Kuchenfresser has decided to put behind him the harrowing, near-drowning experience of Chap. 5. He's invited Ms. Banker to join him for an outing. It's Saturday, and they plan to hike to the top of Mr. Kuchenfresser's favorite nearby peak for a picnic. In addition to everything else, they might need for the day, they're planning to pack a knapsack with enough food and water to sustain them during their trek.

Filling a Knapsack

Let's define our hikers' problem as filling a knapsack of fixed capacity, which we here define in terms of weight. Let's imagine that three items are available to choose from with weights and benefits as defined in Table 11.1.

If the available capacity of the knapsack is 10 pounds, then aside from the obvious challenge of eating 5-pound bars of chocolate, how should the knapsack be packed? What will be the benefit associated with the optimal packing decision?

Knapsack problems are classic examples of problems that can be easily solved using dynamic programming (DP). The example here is adapted from Winston (1991), and is an example of a general class of problems in which units of a resource must be allocated among activities to maximize total benefits, subject to a resource constraint. Although the problem may look familiar to those with experience using **linear programming (LP)**, and could be solved as an LP, the solution method of interest here is considerably more flexible than LP.

To illustrate how dynamic programming works, observe that this problem is simple enough that it can be easily solved by enumerating all feasible loading strategies. The goal here is to solve the problem using an approach that can be generalized to larger problems, but in a setup simple enough that we can easily

© The Author(s), under exclusive license to Springer Nature Switzerland AG 2025
G. Shively, *A Beginner's Guide to Dynamic Optimization in Economics*, Classroom Companion: Economics,
https://doi.org/10.1007/978-3-032-09374-5_11

Table 11.1 Provisions for the hike

Item	Weight	Benefit
Water bottle	4 pounds	11
Bag of gorp[1]	3 pounds	7
Chocolate	5 pounds	12

Table 11.2 Feasible knapsack packing options

Item combination	Total weight	Total benefit
2 water bottles	8 pounds	22
3 bags of gorp	9 pounds	21
2 bars of chocolate	10 pounds	24
1 water bottle + 1 bar of chocolate	9 pounds	23
1 water bottle + 2 bags of gorp	10 pounds	25

confirm we've found the solution. The idea is that we can then reliably use the overall approach to solve more difficult problems.

Following the logic introduced in Chap. 1, we are interested in a feasible solution path. The relevant feasible options available are listed in Table 11.2.

It is clear from some quick addition and a glance at the entries in Table 11.2 that a packing approach that includes all three items is not feasible, and the best available option is to pack one water bottle and two bags of gorp. This provides a total benefit of 25. (What, exactly, this benefit represents remains a bit vague, but need not trouble us).

We now want to approach this problem using dynamic programming, the third method available for solving dynamic optimization problems. DP is conceptually simple and can be applied to a wide range of empirical problems in economics. It is applicable whenever decisions are made in stages. It is especially well suited to situations in which functions are discontinuous, where decision variables take on discrete values (as in our knapsack problem) or where randomness and uncertainty prevail.

Dynamic programming was developed in the 1950s by Richard Bellman. It can be applied to both discrete-time and continuous-time problems. The basic principle of dynamic programming is the **principle of optimality**. In Bellman's words:

> An optimal policy has the property that, whatever the initial state and optimal first decision may be, the remaining decisions constitute an optimal policy with regard to the state resulting from the first decision (R. Bellman, 1957, p. 83).

---

[1] Some readers may be unfamiliar with the term *gorp*, which refers to a high-calorie mixture of snacks designed to provide maximum energy for a given unit of weight. Although often identified as an acronym for "Good Old Raisins and Peanuts," linguists have established that the word predates this definition, making it an after-the-fact **backronym**. These days, the term gorp seems to have fallen out of use, having been replaced by the less interesting but more informative supermarket label *trail mix*.

In other words, if one uses the state value that results from early decisions as the initial state for subsequent decisions, then the controls over the remaining periods are certain to be optimal, conditional on the early decisions. This, it turns out, is a very powerful result.

Bellman's Equation

The central tool used in dynamic programming is **Bellman's equation**. To derive Bellman's equation, recall our basic optimal control problem:

$$V[x] = Max \int_0^T e^{-rt} F(x(t), u(t), t) dt \tag{11.1}$$

subject to:

$$\dot{x} = g(x(t), u(t), t)$$

$$x(0) = x_0$$

$$x(T) = x_T.$$

Now, consider breaking this problem into two pieces. Call these pieces Part A and Part B. We'll let Part A cover the time interval $[0, s]$ and let Part B cover the remaining time interval $[s, T]$. Bellman's principle of optimality says that choosing a control u over the interval $[0, s]$ is the same as choosing a control u over the interval $[0, T]$, with the understanding that after time s, optimal decisions will be made conditional on the state that prevails at time $t = s$, i.e., x_s.

To see this, start with the second problem, Problem B (i.e., $s = 0$):

$$V_s[x_s] = Max \int_s^T e^{-rt} F(x(t), u(t), t) dt \tag{11.2}$$

subject to:

$$\dot{x} = g(x(t), u(t), t)$$

$$x(s) = x_s.$$

The principle of optimality says that the original problem (from 0 to T) can be restated as follows:

$$V_0[x_0] = Max \int_0^s e^{-rt} F(x(t), u(t), t)dt + e^{-rs} V_s[x_s] \qquad (11.3)$$

subject to:

$$\dot{x} = g(x(t), u(t), t)$$

$$x(0) = x_0.$$

Notice that $V_s[x_s]$ is not inside the integral of Eq. (11.3), but is instead a scalar value added at the end. In this case, $V_s[x_s]$ carries all of the information about what will happen beyond s, and the assumption is made that decisions from s to T will be made optimally. In other words, one can solve an initial problem, putting aside considerations regarding future decisions, secure in the knowledge that future decisions will be made optimally. If this sounds a bit confusing, hold on. Let's take Bellman's approach to solving our knapsack-filling problem.

First, although our problem is somewhat trivial, we want to approach it formally to lay out the general approach. All DP problems begin with a careful definition of the problem. So, let us use i as an index over items and identify items by x. In other words, x_i is the number of items of type i to be considered. In this problem, $i =$ {water, gorp, and chocolate}. We also need to define stages. In doing so, we let each stage represent a decision about whether or not to add an item to the knapsack. It is completely arbitrary which items we associate with which stages. Below, we will work backward through our list, starting with chocolate and ending with water. Using t to note the stage, therefore, we have $t = 3$ corresponding to chocolate, $t = 2$ corresponding to gorp, and $t = 1$ corresponding to water.

Now, let's define the benefit associated with an item as r (for reward) and make it a function of the number of items of that type being considered. These rewards are listed in Table 11.1. In this problem, x_1 is the number of water bottles under consideration and $r_1(x_1) = 11x_1$ is the benefit of adding water to the pack; $r_2(x_2) = 7x_2$ is the benefit of adding gorp to the pack; and $r_3(x_3) = 12x_3$ is the benefit of adding chocolate to the pack. One bottle gives a benefit of 11, two bottles give a benefit of 22, etc. Note that here we assume marginal utility is constant, which means $r(x)$ is simply a multiplication of the number of items times its unit weight. In practice, the benefit of an additional unit may decline as additional units are added. We could easily accommodate diminishing marginal utility by altering the form of $r(x)$.

Let's represent the cost associated with adding units by $g(x)$. So, $g(x_1) = 4x_1$ says that each additional unit of item 1 (water) uses 4 units of our resource (in this case, capacity measured in terms of weight), $g(x_2) = 3x_2$ says that each additional unit of item 2 (gorp) uses 3 units of our resource, and $g(x_3) = 5x_3$ says that each additional unit of item 3 (chocolate) uses 5 units of our resource. One unit of gorp uses 3 pounds, 2 units of gorp use 6 pounds, etc.

Now define $f_t(d)$ as the maximum benefit possible from a d-pound knapsack filled with items of type 1, 2, and 3. We have the necessary data in the form of:

$$r_1(x_1) = 11x_1 \qquad g(x_1) = 4x_1$$
$$r_2(x_2) = \ 7x_2 \qquad g(x_2) = 3x_2$$
$$r_3(x_3) = 12x_3 \qquad g(x_3) = 5x_3.$$

Now we can write out the possible configurations for the backpack. The goal is to do this systematically and somewhat mechanically. So, as we laid out earlier, stage 3 represents the act of adding chocolate; stage 2 represents the act of adding gorp; and stage 1 represents the act of adding water. We start with the final stage and solve the problem backwards, enumerating all feasible combinations at each stage. This part is tedious, but once we have the basic pattern down, it goes pretty quickly.

Stage 3: Considering Chocolate

$$f_3(d) = max\{12x_3\}$$
$$f_3(10) = 24$$
$$f_3(5) = f_3(6) = f_3(7) = f_3(8) = f_3(9) = 12$$
$$f_3(0) = f_3(1) = f_3(2) = f_3(3) = f_3(4) = 0$$
$$x_3(10) = 2$$
$$x_3(5) = x_3(6) = x_3(7) = x_3(8) = x_3(9) = 1$$
$$x_3(0) = x_3(1) = x_3(2) = x_3(3) = x_3(4) = 0$$

These entries simply indicate that a knapsack with a remaining capacity of 10 pounds can be maxed out with two bars of chocolate, a knapsack with a remaining capacity of 5–9 pounds can accommodate 1 bar of chocolate, and a knapsack with a remaining capacity of 0–4 pounds can accommodate no bars of chocolate. The important thing to note here is that we don't ask how we got to the state (remaining capacity) we are in. We simply ask what we can do with what we are given.

Now, let's back up to the next stage of the problem.

Stage 2: Considering Gorp

Bellman's equation tells us that:

$$f_2(d) = max\{7x_2 + f_3(d - 3x_2)\}$$

where $3x_2 \le d$ must hold.

In words, we seek to maximize the combination of some amount of gorp along with whatever implication this choice has for us in the next stage of the problem, when we consider adding chocolate. By choosing some amount of gorp, we necessarily impose a feasible set of choices in stage 3. More gorp probably means less chocolate; less gorp, more chocolate. The cleverness here is that we don't worry about stage 3, we only assume that once we get there, we do the best we can with what we have been given.

From this, we get:

$$f_2(10) = max \begin{cases} 7(0) + f_3(10) = 24^* & x_2 = 0 \\ 7(1) + f_3(7) = 19 & x_2 = 1 \\ 7(2) + f_3(4) = 14 & x_2 = 2 \\ 7(3) + f_3(1) = 21 & x_2 = 3 \end{cases}$$

so that $f_2(10) = 24$ and $x_2(10) = 0$. Note that the maximum value for $f_2(10) = 24$, which is marked with an asterisk (*). We repeat the exercise for $f_2(9)$ and so forth to obtain:

$$f_2(9) = max \begin{cases} 7(0) + f_3(9) = 12 & x_2 = 0 \\ 7(1) + f_3(6) = 19 & x_2 = 1 \\ 7(2) + f_3(3) = 14 & x_2 = 2 \\ 7(3) + f_3(0) = 21^* & x_2 = 3 \end{cases}$$

so that $f_2(9) = 21$ and $x_2(9) = 3$.

$$f_2(8) = max \begin{cases} 7(0) + f_3(8) = 12 & x_2 = 0 \\ 7(1) + f_3(5) = 19^* & x_2 = 1 \\ 7(2) + f_3(2) = 14 & x_2 = 2 \end{cases}$$

so that $f_2(8) = 19$ and $x_2(8) = 1$.

$$f_2(7) = max \begin{cases} 7(0) + f_3(7) = 12 & x_2 = 0 \\ 7(1) + f_3(4) = 7 & x_2 = 1 \\ 7(2) + f_3(1) = 14^* & x_2 = 2 \end{cases}$$

so that $f_2(7) = 14$ and $x_2(7) = 2$.

$$f_2(6) = max \begin{cases} 7(0) + f_3(6) = 12 & x_2 = 0 \\ 7(1) + f_3(3) = 7 & x_2 = 1 \\ 7(2) + f_3(0) = 14^* & x_2 = 2 \end{cases}$$

so that $f_2(6) = 14$ and $x_2(6) = 2$.

$$f_2(5) = max \begin{cases} 7(0) + f_3(5) = 12^* & x_2 = 0 \\ 7(1) + f_3(2) = 7 & x_2 = 1 \end{cases}$$

so that $f_2(5) = 12$ and $x_2(5) = 0$.

$$f_2(4) = max \begin{cases} 7(0) + f_3(4) = 0 & x_2 = 0 \\ 7(1) + f_3(1) = 7^* & x_2 = 1 \end{cases}$$

so that $f_2(4) = 7$ and $x_2(4) = 1$.

$$f_2(3) = max \begin{cases} 7(0) + f_3(3) = 0 & x_2 = 0 \\ 7(1) + f_3(0) = 7^* & x_2 = 1 \end{cases}$$

so that $f_2(3) = 7$ and $x_2(3) = 1$.

$$f_2(2) = 7(0) + f_3(2) = 0 \; x_2 = 0$$

so that $f_2(2) = 0$ and $x_2(2) = 0$.

$$f_2(1) = 7(0) + f_3(1) = 0 \; x_2 = 0$$

so that $f_2(1) = 0$ and $x_2(1) = 0$.

$$f_2(0) = 7(0) + f_3(0) = 0 \; x_2 = 0$$

so that $f_2(0) = 0$ and $x_2(0) = 0$.

Note that the number of feasible choices declines rapidly as we work through the possible states for stage 2 of the problem. We now move on to Stage 1, which is trivial.

Stage 1: Considering Water

$$f_1(10) = max \begin{cases} 11(0) + f_2(10) = 24 & x_1 = 0 \\ 11(1) + f_2(6) = 25^* & x_1 = 1 \\ 11(2) + f_2(2) = 22 & x_1 = 2 \end{cases}$$

At the start, we have 10 pounds at our disposal. We can feasibly do one of three things: use none of it for water and keep the original capacity of the knapsack, pack one bottle of water and give up some of the capacity, or use all of the capacity of the knapsack for water (two bottles). If we pack no water, we derive no benefit in stage 1, but we move to the next stage with 10 units of capacity and know that with that amount of weight remaining, we can achieve a total benefit of 24. From the perspective of stage 1, it doesn't matter how we get that amount of benefit in stage 2. We only need to know what the feasible maximum is. In contrast, if we choose to pack two water bottles, we obtain a benefit of 22 and pass into the subsequent stage with 0 available weight remaining, which cannot be converted into subsequent benefits. We find that the maximum benefit to be gained from filling a 10-pound knapsack comes from filling it with one type-1 item and then accepting the return associated with whatever choices are optimal from that point forward. The implication is that we will subsequently choose two type-2 items. This results in a total benefit of 25, which is what we were expecting based on our initial, full-enumeration analysis.

Recursion

What's so special about this approach? The answer is that it is **recursive**. The recursive nature of the problem means that we assume the last part of the problem has already been solved when we optimize the first part of the problem. The connection between the principle of optimality and this process of recursion is an essential characteristic of DP and gives rise to the concept of backward recursion, which was introduced briefly in Chap. 1. Backward recursion simply means solving problems backward. The classic statement and outline of recursive methods can be found in the book by Bellman and Dreyfus (1962). The problem outlined in Eq. (11.3) above suggests that, even if we somehow behaved sub-optimally in the interval $[0, s]$, as long as we execute the optimal decision for the interval $[s, T]$, we can receive the amount $e^{-rs}V_s[x_s]$, which will be—by construction—optimal.

The recursive nature of dynamic programming means that DP can be viewed in two ways. It is an *approach* to solving problems, and it is a *method* for solving problems.

As a **solution approach,** dynamic programming is a method for analyzing problems that explicitly recognize the principle of optimality and express the maximization problem as the following functional equation:

$$V[x] = Max[F(x, u) + \beta V[z(x, u)]]. \tag{11.4}$$

In Eq. (11.4), β is a discount factor and z is a **transformation function** that describes the value of x in the next period, conditional on the current period decision. The z function plays the role of the equation of motion from optimal control theory and is sometimes referred to as a **return function**, since it tells us something either about the return (i.e., benefit) associated with a decision, or the state to which the system returns in the subsequent period. In the knapsack problem, the return function tells us what we will get when we pass to the next stage of the problem with a given amount of capacity (weight) remaining. Equation 11.4 is the Bellman equation and $V[x]$ is the value function.

As a **solution method,** dynamic programming is a practical method of solving problems using recursive methods, either by pencil and paper (as in the knapsack problem) or with the aid of a computer. Rust (2018) reviews applications to economics.

In general, dynamic programming makes few technical demands, certainly fewer than the calculus of variations or optimal control theory, especially in terms of requirements on admissible functions, controls, etc. Note, however, that the DP algorithm generally requires that the problem to be solved possess two properties. These are:

Separability: for any time period, the functions must depend on the current values of the state and control variables but not on their past or future values; and

Additivity: the total benefit must equal the sum of net benefits across stages.

Example: Finding the Least-Cost Route of Travel

Now that we've worked through a knapsack problem using dynamic programming, let's try something different. Recall the network travel example of Chap. 1, as illustrated in Fig. 1.1. We now want to revisit the example and solve it again, not by using the brute force method of listing all possible paths, but by using backward recursion. Recall that a hypothetical traveler wants to travel from location A to location J using the least-cost route. We identify transit points along the route by circles A, B, C, D, E, F, G, H, I, and J, and we identify the sub-cost of moving between pairs of nodes by the values on the arcs connecting the nodes. To find the minimum-cost path from the starting node to the final node, we develop Bellman's equation as a series of **cost-to-go** functions. These take the form:

$$V[i] = Min[c_{ij} + V(j)]]\tag{11.5}$$

where $V[i]$ is the cost from node i to the final node, c_{ij} is the cost of moving between node i and node j, and it is understood that we are searching across the set of all nodes that are reachable from node i. We want to start at the final stage and final node and recursively work backward, calculating Eq. (11.5) at each node. We'll repeat the recursion until we reach the starting node, at which time Eq. (11.5) will reveal the minimum cost. As a side-project, we'll need to keep track of paths as we go, so we know which is the least-cost path when we find it.

Starting with the final node, J, we have:

$$V[J] = 0$$

Since, it is clear that there is no travel beyond J, and hence no cost associated with continuing beyond the fourth stage of the problem.

Now, let's back up to stage 3, the penultimate stage. We have three options:

$$V[G] = c_{GJ} + V(J) = 6 + 0 = 6$$
$$V[H] = c_{HJ} + V(J) = 3 + 0 = 3$$
$$V[I] = c_{IJ} + V(J) = 4 + 0 = 4.$$

This tells us what to expect when we find our traveler at various nodes in stage 3 of the problem. We can now use this information to examine stage 2, which is where the approach gets interesting:

$$V[D] = c_{DG} + V(G) = 4 + 6 = 10$$

$$V[E] = Min[c_{EG} + V(G), c_{EH} + V(H)] = Min[4 + 6, 3 + 3] = Min[10, 6] = 6$$

$$V[F] = Min[c_{FH} + V(H), c_{FI} + V(I)] = Min[5 + 3, \ 4 + 4] = Min[8, 8] = 8.$$

As we can see for $V[E]$ and $V[F]$, we implicitly account for the fact that, at these nodes, our traveler will face a choice: E→G or E→H when at E, or F→H or F→I when at F. By identifying the optimal decision at this stage, and knowing that we will choose it when we come to it, we can now back up to the preceding stage, knowing that the path forward will be the optimal one. In this particular example, the information provided is somewhat trivial, since there is only one pair of choices at each node, but imagine if there were hundreds of options available. By identifying the optimal decision at this stage, all we need to carry with us in the next step backward is the confidence that we will pick the correct path when we get to it, without enumerating every possible path that follows. This represents a tremendous savings of time, effort, calculation, or computing.

Now let's follow the same approach and back up to stage 2, which consists of two nodes:

$$V[B] = Min[c_{BD} + V(D), c_{BE} + V(E)] = Min[3 + 10, 6 + 6] = Min[13, 12] = 12$$

$$V[C] = Min[c_{CE} + V(E), c_{CF} + V(F)] = Min[3 + 6, 7 + 8] = Min[9, 15] = 9.$$

Now, at the start of the problem, we find:

$$V[A] = Min[c_{AB} + V(B), c_{AC} + V(C)] = Min[4 + 12, 2 + 9] = Min[16, 11] = 11.$$

The minimum cost from A to J, therefore, is 11 and, tracing back, the optimal path is A→C→E→H→J, which confirms what we found in Chap. 1. For

Table 11.3 Backward recursion results for minimum-cost travel problem

Stage	Node	Next node	Cost to next	Cost-to–go	Total cost	Minimum cost
	J	-	-	-	-	0
4	G	J	6	0	6	6
4	H	J	3	0	3	3
4	I	J	4	0	4	4
3	D	G	4	6	10	10
3	E	G	4	6	10	6
3	E	H	3	3	6	
3	F	H	5	3	8	8
3	F	I	4	4	8	
2	B	D	3	10	13	12
2	B	E	6	6	12	
2	C	E	3	6	9	9
2	C	F	7	8	15	
1	A	B	4	12	16	11
1	A	C	2	9	11	

clarity, the process of backward recursion is summarized in Table 11.3. It should be easy to confirm that this process conforms to Bellman's principle of optimality. We've shown, in essence, that if we trust that at any point in a sequence of decisions, remaining decisions are made optimally with regard to the state resulting from the preceding decision, then whatever the state and optimal decision may be, the backward recursion approach provides the optimal policy.

Resource Allocation Problems

It is instructive to think about dynamic programming in general terms. From the perspective of economics, the value of dynamic programming resides in the fact that DP can be used to solve a wide range of **generalized resource allocation problems**, whether or not they involve time. In a seminal paper, Burt (1962) discusses a number of potential applications.

For example, suppose we have w units of a resource available for T activities. Each activity has costs and benefits associated with implementation. How does one allocate resources to maximize benefit, subject to a constraint on available resources?

To examine this generic style of problem, we need some definitions. Let's build on what we have already used for the knapsack problem.

Let

  w = resource available
  T = number of activities
  x_t = level of implementation of activity t $(x_t \geq 0)$
  $g_t(x_t)$ = units of resource used by activity t
  $r_t(x_t)$ = benefit of activity level

The resource allocation problem is to:

$$Max \sum_{t=1}^{t=T} r_t(x_t)$$

$$s.t. \sum_{t=1}^{t=T} g_t(x_t) \leq w. \tag{11.6}$$

For sake of example, some possible interpretations of r, g, and w, excluding the knapsack problem we've already encountered, are given in Table 11.4.

As we have seen, to solve a resource allocation problem using dynamic programming, one must follow several steps. The steps are reviewed here in detail, as they are instructive for developing computer-based algorithms for solving problems.

First, define $f_t(d)$ as the maximum benefit that can be obtained from activities $t, t+1, \ldots T$ if d units of the resource are allocated to activities $t, t+1, \ldots T$. In this case, d is our control variable. Then, the solution to a resource allocation problem must take the form:

$$f_{T+1}(d_t) = 0 \, \forall d$$

$$f_t(d_t) = max \, \{r_t(x_t) + f_{t+1}(d_t - g_t(x_t))\} \tag{11.7}$$

where x_t is a non-negative integer satisfying $g_t(x_t) \leq d_t$.

Table 11.4 Examples of resource allocation problems

Benefit	Resource requirement	Unit
$r_t(x_t)$	$g_t(x_t)$	w
Sales of a product in a region	Cost of assigning sales reps to the region	Total sales force budget
Value of agricultural production	Land area allocated to individual crops	Total land area available
Value of various items of furniture	Legs, seats, backs, and other components	Board feet of lumber
Product sales	Transportation between warehouses and outlets	Fuel or time
Daily care of sick patients	Nurses and doctors	Available staff

Now that we've solved the knapsack problem, we can better understand what this means.

Consider a problem that ends at time T. The first part of (11.7), the statement $f_{T+1}(d_t) = 0 \, \forall \, d$, simply says that, regardless of the decision taken, no value can be obtained *after* the problem ends. This is important, because it means that we know something about the terminal condition for the problem. On the other hand, the second part of (11.6), namely the statement $f_t(d_t) = \max\{r_t(x_t) + f_{t+1}(d_t - g_t(x_t))\}$ means that the maximum benefit at any time t will be derived from a decision that takes account of two things: the impact of the decision on current benefits $r_t(x_t)$ and the impact of the decision on all subsequent benefits. The important feature of this setting up the problem in this way, as we have seen, is that *all* future benefits are compressed and represented by a single value, $f_{t+1}(d_t - g_t(x_t))$ where $g_t(x_t)$ tells us what the subsequent state will be, conditional on the current decision. We assume that all future decisions will be optimal and therefore that f_{t+1} is maximized.

As we've seen already, an optimal allocation of resources to activities is determined in three steps, which are listed here in the form of our backward recursion:

Step 3: Start at the end of the problem. Determine all $f_T(\cdot)$ and $x_T(\cdot)$.

Step 2: Back up one stage. Use $f_t(d_t)$ to determine all f_{T-1} and x_{T-1}. Continue backing up in this way until all f_2 and x_2 have been determined. Calculate $f_1(d_1)$ and $x_1(d_1)$.

Step 1: Implement activity 1 at level $x_1(w)$, which leaves $w - g_1(x_1(w))$ units of the resource available for activities 2, 3 ... T. Implement activity 2 at a level $x_2[w - g_1(x_1(w))]$. Continue until all activity levels have been determined.

The only way to really get comfortable with DP is to work some examples. It is useful to start with something easy, like our knapsack problem, and then move on to more difficult problems, such as those posed by Bertsekas (1987) or Kennedy (1986). Although the method of solution outlined in our knapsack problem might appear a bit tedious, it is important to point out that, in the process of writing down the steps, we generated an algorithm that can be easily converted to computer code and extended to problems with a much larger state and control space. This is essentially how DP is used in practice, for example, in such practical matters as inventory management or loading container ships.

Optimal Stopping Problems

No introduction to dynamic programming would be complete without reference to what are called **optimal stopping** problems. These are classic problems in mathematics and decision theory in which the goal is to select the best option from an unfolding sequence of options or to choose the best time to act in order to maximize expected benefits or minimize expected cost. These problems appear in many fields, including economics, and are often labeled as **the secretary problem**. The key aspect of such problems is that a decision must be made at each stage regarding whether to stop the process or continue it.[2] Anyone who has ever tried to sell a used car has already solved this problem. The basic solution strategy is relevant to a large number of timing decisions in economics, including investment purchases, asset sales, and option pricing.

The setup is as follows. Imagine a manager who wants to hire a new secretary. Working with her human resources (HR) officer, she drafts the position description. HR posts it, collects and reviews résumés, and creates a short list of equally-qualified (on paper) candidates for the manager to interview in-person. The interviews are scheduled sequentially over the course of one day, with the candidates arriving in random order. At the start, the manager has no preference over the candidates and only knows that they have been deemed minimally qualified based on the assessment of the firm's HR office. The somewhat unnatural twist that renders this problem interesting to us is that the decision about each particular applicant must be made immediately at the end of the interview. Once rejected, applicants cannot be recalled. During the interview, the administrator can rank applicants among all applicants interviewed thus far, but has no information on the quality of applicants not yet interviewed.

One way to visualize the secretary problem is by laying out a simulation, say of 20 random interviews, as in Fig. 11.1. In this illustration, none of the first seven candidates are selected. Instead, they are used to establish a baseline score to which the remaining 13 candidates are compared. Note that there is no guarantee in these problems that we select the best candidate, only that we maximize the expected value, for example, our chances of selecting the best candidate. Another way to visualize the problem is using a decision tree, as done for $n = 5$ in Fig. 11.2. We'll solve a version of this particular five-candidate problem below.

The secretary problem is frequently solved using probability theory.[3] However, it can also be framed as a **Markov Decision Process (MDP)**. We will explore Markov processes in greater depth in Chap. 12, where we take up the issue of uncertainty in a more formal way. But the secretary problem also can be easily formalized using a Bellman equation, which underscores the recursive decision-making process. The Bellman equation for the secretary problem captures the idea of making an optimal decision at each step based on the expected future reward. Letting P represent the payoff at a stage, the general form is:

---

[2] The secretary problem is similar but different from what is known as the **newsboy problem**, in which an irreversible decision (how many papers to stock) is made at the *start* of the problem, with incomplete information about demand. The newsboy problem is a static optimization problem, whereas the secretary problem, in which the irreversible decision is encountered at every stage, is inherently dynamic because the decision made at any stage influences the options available in subsequent stages.

[3] Using the probabilistic approach, mathematicians have developed optimal stopping rules. One of the most reliable is the $1/e$ stopping rule, which relies on rejecting the first n/e interviewees (where n is the number of interviewees in the pool and e is the base of the natural logarithm). These rejected candidates are used to form a benchmark for selection. The interviewer then hires the first applicant who is better than the benchmark. This strategy is called the $1/e$ strategy because it selects the best candidate approximately $1/2.718$, or 37%, of the time.

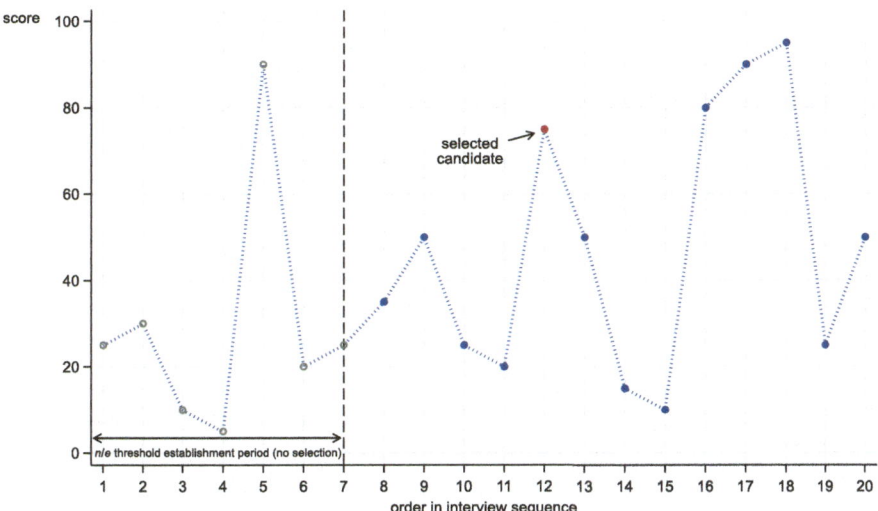

Fig. 11.1 A secretary problem simulation with $n = 20$ job candidates

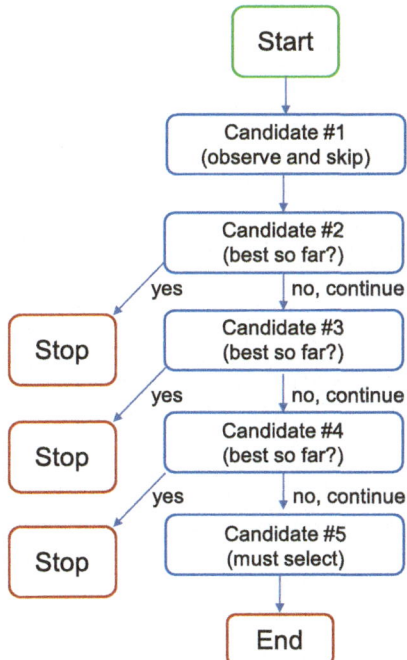

Fig. 11.2 Secretary problem with $n = 5$ job candidates

$$V(i) = max\{P(stopping\ at\ i),\ P(continuing)\}. \qquad (11.8)$$

To be more specific, we can let n be the total number of candidates and i be the number of candidates interviewed thus far. Identify $R(t)$ as the rank of the t–th candidate interviewed and $V(i)$ as the expected maximum quality of a candidate selected from the first i candidates interviewed. The Bellman equation for the problem is:

$$V(i) = max\left\{\frac{t}{i}R(t),\ \left(1 - \frac{t}{i}\right)V(i+1)\right\} \qquad (11.9)$$

where $\frac{t}{i}$ represents the probability of stopping at candidate t.

Let's examine what backward recursion looks like with a pool of five candidates. Following the steps outlined above, and beginning in the final stage, we know that if the manager reaches the final candidate, this applicant must be hired. The probability that this applicant is the best candidate is simply $1/5 = 0.2$. So, we have:

$$V(5) = 0.20$$

Now we back up one step. The manager must choose between candidate 4 and continuing to stage 5, where we know the probability of success is 0.20. So, Bellman's equation tells us the choice is:

$$V(4) = max\left\{\frac{1}{4},\ V(5)\right\}$$

or

$$V(4) = max\{0.25,\ 0.20\} = 0.25.$$

We back up again.

$$V(3) = max\left\{\frac{1}{3},\ V(4)\right\}$$

or

$$V(3) = max\{0.33,\ 0.25\} = 0.33.$$

We back up again.

$$V(2) = max\left\{\frac{1}{2},\ V(3)\right\}$$

or

$$V(2) = max\{0.50, 0.33\} = 0.50.$$

For the first candidate, of course, we have no prior information. A reasonable strategy would be to meet and reject this candidate in order to establish a baseline, accepting the risk that this could be the best candidate in the pool. So, $V(1) = 0.50$. that's the *maximum* probability of success with five candidates. The manager skips the first candidate and then proceeds to pick the first candidate who is better than candidate 1. If the manager reaches candidate 5, then she must pick that candidate regardless of how happy she might be about it.

In general, the success rate for the secretary problem (in terms of getting closest to hiring the "best" candidate) increases as the pool grows larger. This is because the candidates appearing early in the process can be passed over while providing information about the overall pool. With small sample sizes, the information gained from meeting with these initial candidates tends to be outweighed by the randomness of the pool. If the hiring decision could be taken after interviewing all candidates, then it would be easy enough to keep track of the candidate ranking and then select the best candidate at the end. In fact, this is how most hiring works. The challenge associated with the optimal stopping problem is that we insist that the hiring decision be made immediately after interviewing a candidate. This comes closer to mimicking asset selling decisions or other kinds of financial market transactions in which a yes or no decision must be made on the spot and cannot be postponed or reversed.

As a final example of using the DP approach, we end the chapter by re-solving a problem we encountered earlier as an optimal control problem in Chap. 8. We must choose $\{u\}$ for $t = 1, 2, 3$ to maximize:

$$\sum_{t=1}^{3} ln(u_t)$$

subject to:

$$x_{t+1} = 1.25x_t - u_t$$

$$x_1 = 1$$

$$x_4 = 1.25.$$

For this problem, Bellman's equation is:

$$V_t[x_t] = max\{ln(u_t) + V_{t+1}[x_{t+1}]\}.$$

We solve the problem backwards.

Begin in period 3:

$$V_3[x_3] = max\{ln\ (u_3)\}$$

subject to

$$x_4 = 1.25x_3 - u_3$$

$$x_4 = 1.25.$$

There is only one point in the feasible set, so we require $u_3 = 1.25x_3 - 1.25$. This leads to:

$$V_3[x_3] = ln(1.25x_3 - 1.25).$$

The period 2 problem is:

$$V_2[x_2] = max\ \{ln(u_2) + V_3[x_3]\}$$

subject to:

$$x_3 = 1.25x_2 - u_2$$

$$x_2\ given.$$

Substituting for $V_3[x_3]$ yields:

$$V_2[x_2] = max\ \{ln(u_2) + ln(1.25x_3 - 1.25)\}$$

$$= max\ \{ln(u_2) + ln(1.5625x_2 - 1.25u_2 - 1.25)\}.$$

Although we can enumerate all of the possible state/control combinations, we can speed up the process using optimality conditions.[4] The first-order condition $\partial V/\partial u_2$ yields:

$$u_2^* = 0.625x_2 - 0.5.$$

---

[4] We won't go into detail here, but a common approach to solving empirical dynamic programming problems (for example, via computer) is to approximate a large continuous state space by **discretizing** it into a series of pieces and doing a **grid search** across these pieces, looping through finer and finer divisions of the grid until one is confident of having reached convergence on the solution.

When substituted back into the problem, this gives:

$$V_2[x_2] = ln(0.48828) + 2ln(x_2 - 0.8)$$

$$V_1[x_1] = max\{ln(u_1) + V_2[x_2]\}$$

subject to:

$$x_2 = 1.25x_1 - u_1$$

$$x_1 \ given.$$

Substituting for $V_2[x_2]$ and x_2 provides:

$$V_1[x_1] = max\{ln(u_1) + ln(0.48828) + 2 \ ln(1.25x_1 - u_1 - 0.8)\}.$$

The first order condition $\partial V/\partial u_1$ yields:

$$u_1^* = 0.4167x_1 - 0.2667.$$

Now we can substitute for $x_1 = 1$. Work forward in time to get:

$$u_1 = 0.15$$

$$x_2 = 1.25 - 0.15 = 1.1$$

$$u_2 = 0.625(1.1) - 0.5 = 0.1875$$

$$x_3 = 1.25(1.1) - 0.1875 = 1.1875$$

$$u_3 = 1.25x_3 - 1.25$$
$$= 1.25(1.1875) - 1.25$$
$$= 0.234375.$$

A quick comparison will show that this matches what we found via optimal control in Chap. 8.

References

Bellman, R. (1957). *Dynamic programming*. Princeton University Press.
Bellman, R., & Dreyfus, S. (1962). *Applied dynamic programming*. Princeton University Press.

Bertsekas, D. P. (1987). *Dynamic programming: Deterministic and stochastic models.* Prentice-Hall, Inc.

Burt, O. R. (1962). Dynamic programming: Has its day arrived? *Western Journal of Agricultural Economics, 7*(2), 381–393.

Kennedy, J. O. S. (1986). *Dynamic programming: Applications to agriculture and natural resources.* Elsevier Applied Science Publishers.

Rust, J. (2018). Dynamic programming. In *The new Palgrave dictionary of economics.* Palgrave Macmillan.

Winston, W. L. (1991). *Operations research: Applications and algorithms.* PWS-Kent Co.

A World of Uncertainty

12

Uncertainty is all around us and dealing with it is a fact of life (and economics).

Playing the Game *Monopoly*

Aside from the ancient games of chess, checkers and *Go*, the most popular board game of all time is most likely *Monopoly*.[1] For those not familiar with this multi-player game, it involves a group of players taking turns rolling a pair of dice and moving a curiously old-fashioned playing piece around a square board. The board consists of 40 spaces identified (in most cases) as fictitious properties of varying value. The goal of the game is to acquire these properties and the payment streams conferred by ownership. Payments, which can become quite onerous, must be made by opposing players who land on owned property. The ultimate in competitive games, the winner is the one who bankrupts all of the other players. Playing it can be a rather nasty experience and can quickly turn friends into frenemies.

Putting these features aside for the moment, an interesting aspect of the game from our perspective is that it involves a tremendous amount of repetition—as players take turns rolling dice and circling the board—laid atop a great deal of uncertainty. The game takes a long time to play, in some cases many hours or even days. During a game, the fortunes of players can swing wildly and near the end of the game, the suspense surrounding

---

[1] According to reporting by Leopold (2015) (just search for the CNN article "Monopoly at 80"), the game has been printed in more than 46 languages and has sold more than 275 million copies worldwide. Properties in the game are named after locations in and around Atlantic City, New Jersey. The history of *Monopoly* can be traced back to 1903, when (somewhat ironically) an American *anti-monopolist*, Lizzie Magie, created *The Landlord's Game* to explain the single-tax theory of the American economist Henry George, which is described in his 1881 book *Progress and Poverty* (George, 1881).

© The Author(s), under exclusive license to Springer Nature Switzerland AG 2025
G. Shively, *A Beginner's Guide to Dynamic Optimization in Economics*, Classroom Companion: Economics,
https://doi.org/10.1007/978-3-032-09374-5_12

players' rolls of the dice can add a considerable element of excitement (or dread). There is so much uncertainty embedded in the game, and so much of the game's dynamics are governed by the behaviors of the various players, that finding an optimal strategy to solve what is, in essence, a dynamic optimization problem, is nearly impossible. However, some strategies seem to work better than others. Why might that be the case?

At root, *Monopoly* is governed by a **Markov process**. A Markov process is a stochastic process in which the probability distribution associated with reaching a particular state at time $t + 1$ depends only on the state at time t and not on any of the states that were passed through on the way to reaching that state. The process is said to be a **memoryless process**: the only history that matters is current history, and all of the information needed to understand what might potentially happen in the future is embodied in the current state.

In the case of *Monopoly*, we can define each player's position on the board as their current state. We can also define each player's turn as a stage in the problem. One's probability of landing on any given property (i.e., moving from an existing state to a different state) depends only on one's current location (the current state) and the roll of the dice. At each turn (stage) one moves from the current location on the board (state) to a position from 2 to 12 spaces further along the square, looping around the board in a continuous fashion until one either runs out of money and exits the game or acquires it all and wins. Importantly, at each stage the transition probability is independent of stage and state; it depends only on one's current position and not on previous stages of play. In other words, if one is sitting on the property known as Indiana Avenue, exactly how one reached that location is irrelevant. It could be by rolling a 7 while sitting on St. James Place, or by rolling a 3 from Free Parking, or by rolling two pairs of double sixes in succession starting from Boardwalk. In other words, all of the spaces that one landed on or passed through in reaching the current location—the past states—are irrelevant. Similarly, it doesn't matter whether you've acquired Boardwalk by buying it, trading for it or receiving it as part of some deal worked out with another player. The fact that you own it at a point in time is all that matters. As described, the game follows a Markov process.

If you are a fan of *Monopoly* then you will immediately see the need to complicate the analysis to characterize the game completely. So, for example, additional state variables might include the number of owned and developed properties on the board, cash on hand or the number of players. An insightful pair of articles on the game appeared in the April and October 1996 issues of *Scientific American* (Stewart, 1996). The author of those articles applies the concept of Markov processes to *Monopoly*, concluding (as many of us have) that some properties on the board are more lucrative to own than others, but that these are not necessarily the ones with the highest rents. They are more likely to be those that have the highest probability of being landed on.

Markov Processes

Markov processes play an important role in dynamic programming, particularly in problems involving decision-making over time under uncertainty. As we have seen, dynamic programming problems are framed in terms of states, where each state represents a particular situation or condition. Markov processes provide a way to model the evolution of states over time probabilistically. So, if movement between states is for some reason uncertain (but perhaps influenced by decisions), and the value function in a problem is therefore an expected but uncertain value, Markov processes provide the stochastic framework needed to model uncertainties and make decisions that minimize expected costs or maximize expected benefits. A good example of a state variable subject to uncertainty might be water in a reservoir. The release of water at critical times of need may be controllable, and hence future states may be influenced by decisions about how much water to release, but the amount of water captured by the reservoir, either as rainfall or snowmelt, is uncertain.

Although one might tend to think about uncertainty as encompassing the great void of what is not known, from an operational perspective it is much easier to confine our imagination and stick to situations in which potential outcomes can be identified and their probabilities of occurrence can be assigned: in other words, the world of known unknowns. Keep in mind that, throughout the discussion that follows, we must assume that there are no gaps in our understanding of what could happen and how likely those outcomes might be. One way of thinking about the shape of uncertainty is to think about what happens when we receive some kind of **disturbance** from the status quo, i.e., a departure from what is expected. Consider something like temperature or rainfall. Although we cannot know for certain whether there will be a heat wave or a drought, or how severe they might be, if we can form reliable guesses about their possibilities then we can enumerate these potential states of the world and assign probabilities to them. We might then begin to take actions to prepare in advance in a way that aligns our best guesses about the distribution of outcomes with our best possible responses. For our purposes, we need to make sure the disturbance space is **countable**, which is another way of saying we can identify and enumerate all of the possible outcomes of interest. If that weren't the case, it would be nearly impossible to conclude anything about how to act in anticipation of a shock.

A special type of problem that appears repeatedly in stochastic dynamic programming is the control of finite and countable state **Markov chains**. As indicated above, a Markov process is one in which the probability distribution associated with reaching a particular state at time $t + 1$ depends only on the state at time t and not on any of the states that were passed through on the way to reaching that state at time t. The mathematical way of stating this is to say a discrete-time stochastic process is a Markov chain if, for all stages $t = 0, 1, 2, \ldots$, and all states: x_0, x_1, $x_2, \ldots$, the following:

$$p(x_{t+1} = k_{t+1} | x_t = k_t, x_{t-1} = k_{t-1}, \ldots x_{t-m} = k_{t-m}, \ldots x_0 = k_0,) \qquad (12.1)$$

is equivalent to:

$$p(x_{t+1} = k_{t+1} | x_t = k_t). \qquad (12.2)$$

In other words, the probability of the state variable in the next stage of the problem, i.e., x_{t+1}, taking on the specific value k_{t+1} conditional on the entire history of values of x, can be reduced to the probability of the state variable in the next stage of the problem taking on the specific value k_{t+1}, conditional on *only* the previous observed value for the state variable. This is what is meant mathematically by the process having no memory.

Key Definitions and Concepts

To go further, we need some additional definitions and concepts. We begin by defining p_{ij} as a **transition probability** from state i to state j. This is simply the probability of moving from one state to another. It may depend on decisions or choices, but we generally make the assumption that for all states the transition probability is independent of time. This leads to what is called a **stationary Markov chain**. This assumption of stationarity is analogous to the concept of autonomous equations we encountered in Chap. 9, i.e., those equations whose functional forms do not change over time. In this case, we say the probability distribution is constant across time.

In most problems, we report transition probabilities in matrix form, i.e., as a **transition probability matrix**. The entries in such a matrix are, by construction, non-negative. Rows correspond to current states (the i's in p_{ij}) and columns correspond to subsequent states (the j's in p_{ij}). Entries tell us the probability of moving from i to j. Values appearing across the rows must sum to 1, since, in essence, some state must be occupied at all times. A typical transition probability matrix looks like this:

$$\mathbf{P} = \begin{bmatrix} 0.10 & 0.20 & 0.50 & 0.20 \\ 0.05 & 0.25 & 0.50 & 0.30 \\ 0.50 & 0.20 & 0.15 & 0.15 \\ 0.05 & 0.50 & 0.05 & 0.40 \end{bmatrix}.$$

Picking a row and a column from $\mathbf{P}$, say row three and column two, the matrix tells us that if, in a particular stage, we find ourselves in state three, the probability of entering state two is 0.20. It is more likely, according to $\mathbf{P}$, that from state three we will move to state one, and it is less likely that we will remain in state three or move to state four. In *Monopoly*, the probabilities associated with moving between locations on the board are defined entirely by the roll of a pair of six-sided dice.

Given two states, i and j, the path from i to j is a sequence of transitions that begin in i and end in j. This complete transition could take several stages. Recall Fig. 1.1. Although the travel problem was not defined in terms of uncertainty, we might now reimagine the various nodes as representing states, and the paths leading between nodes at each stage as being chosen at random, say by a flip of a coin or some other process. In that case, we might be interested in finding not the minimum cost of travel, but determining the minimum *expected* cost of travel.

One always needs to be careful and precise when using transition probabilities to model stochastic dynamic problems. It is therefore worth keeping in mind some specific definitions. First, we say (somewhat obviously) that a state j is **reachable** from a state i if there is a path leading from i to j. In Fig. 1.1, you can confirm that node I is not reachable from node B because there is no path leading from B to I (without backtracking, which we ruled out). We also say that two states i and j **communicate** if i is reachable from j and j is reachable from i.

A set of states is said to be **closed** if no states outside that set are reachable from any state in it. Another way of thinking of this is that once we enter a closed set, we cannot leave the set. It's a bit of a stretch as an example, but imagine two sets of roads: those connecting cities in North America and those connecting cities in Europe. Both belong to the larger set we might call roads, but each is a closed set with respect to the other.

We can also define a state as an **absorbing state** if $\rho_{ii} = 1$. In other words, once the system enters an absorbing state it can never leave the state. For example, in life, and in economic models that intend to approximate the life cycle, such as the **overlapping generations (OLG) model**,[2] death is an absorbing state. In *Monopoly*, exiting the game is an absorbing state, since once you are out of the game, you cannot reenter.[3] It probably goes without saying that an absorbing state is closed!

Another standard classification is to define a state i as a **transient state** if j is reachable from i but i is not reachable from j. For example, we might say high school is a transient state on the way to college, a B.S. is a transient state on the way to a Master's degree, and an M.S. is a transient state on the way to a Ph.D. A more accurate real-world system that exhibits transient states is a bank queue or a customer support call center, since once a customer is served, they leave the waiting state and (hopefully) don't return. In *Monopoly*, being "In Jail" is a transient state because while you might possibly return to jail at some point in the game, you cannot stay there forever—you are forced according to the rules of the game to leave after rolling doubles, using a "Get Out of Jail Free" card, or paying

---

[2] The OLG model is central to the modern study of macroeconomic dynamics and economic growth. In contrast to early macro models in which economic agents live forever, in the OLG model individuals have finite lives that are just long enough to overlap with another agent's life, which opens up opportunities for studying, among other things, bequests and Social Security programs.

[3] In real life, however, bankruptcy for a firm may not be an absorbing state since a firm can sometimes reach a deal with its creditors and emerge from bankruptcy and continue operations, albeit in an altered form.

a fine. By definition, after a large number of periods the probability of being in any transient state is zero. This is because each time we enter a transient state there is a positive probability that we will leave the state and never return.

In contrast, a non-transient state is called a **recurrent state**. For those familiar with the Harold Ramis movie *Groundhog Day*, we might say that Bill Murray's character Phil seems to be stuck in a recurrent state, always starting a new day just like the day before. Fortunately, he manages to find a way to escape his frustrating loop and find a happy ending with Andie MacDowell's character Rita (spoiler alert: love is involved).

A stochastic process that appears on first glance to involve transient states is the weather, since sunny days follow cloudy days. But the weather has recurrent states not, strictly speaking, transient states: in a transient state, there's a chance of exiting and never returning. But if you don't like the weather, you can always just wait for it to improve, which it eventually will.

A state i is said to be **periodic** if all paths leading from i eventually return to i. The length of the path is the period. The Markov chain with the following transition matrix:

$$\mathbf{P} = \begin{bmatrix} 0 & 1 & 0 & 0 \\ 0 & 0 & 1 & 0 \\ 0 & 0 & 0 & 1 \\ 1 & 0 & 0 & 0 \end{bmatrix}$$

is periodic with a period of four. No matter what state we begin in, we will be led back to that same state four periods later. If we initially begin in state 1 (indicated by the first row of column one), with probability 1 we will enter the next period (column two) in state 2. From there, we will move to state 3 in the subsequent stage, and to state 4 in the stage that follows. A return from state 4 to state 1 starts the cycle over again, ad infinitum. A number of biological processes are cyclical in this way: spring follows winter; summer follows spring; fall follows summer; and winter follows fall. Amazingly, some North American cicada emerge in periodic cycles.[4]

Most of the time, economists are interested in dynamic systems that are characterized by non-periodic states and phenomena. A non-periodic state is called **aperiodic**. Economies, in general, are aperiodic. They regularly go through business cycles, swinging from boom to bust, or from periods of growth into recessions, but these do not occur with the kind of regularity that makes them periodic, in the Markov sense. Periods of economic growth and recession are aperiodic. Indeed, a major mandate of policy makers, especially the central banks of most

---

[4] Periodic North American cicadas (often incorrectly referred to as locusts) spend roughly 99% of their lives underground as immature nymphs feeding on fluids from the roots of trees. In their 13th or 17th year of living beneath the surface, cicada nymphs synchronously emerge in late spring, sometimes in impressive numbers.

modern economies, is to ensure that recessions, and the unemployment they entail, are infrequent and short-lived.

If all states in a Markov chain are recurrent, aperiodic, and communicate, then the Markov chain is said to be **ergodic** or to possess **ergodicity**. Ergodic chains and the transition probabilities that characterize them are particularly important in **stochastic dynamic programming**.

Transition Probabilities

It is possible via elementary statistical theory to forecast forward and solve for transition probabilities associated with a Markov process after some passage of time. A straightforward way to calculate the resultant transition probabilities after some given number of stages, say n, is by repeated multiplication of the transition probability matrix. The logic that underlies this process is an underpinning of infinite-horizon dynamic optimization. It relies on the fact that if future states are uncertain, but possible, we can assign long-run probabilities of being in each. If each of those states have, for example, values associated with them (e.g., profits for a firm, utility for an individual, or total consumption for an economy) then we can assess the relative merits of taking steps to steer toward each of these states.

To expand on this idea, consider that infinite-horizon problems are concerned with strategies and solutions that are stable *in the long run*, i.e., regardless of the stage of the problem. If a firm has an infinite horizon then aiming for the best long-run strategy will be optimal since the best long-run strategy will be the one that performs best, independent of stage. Understanding the link between transition probabilities and infinite-horizon dynamic problems is crucial to understanding the logic of stochastic dynamic programming. One can reason more formally as follows. If the underlying stochastic process of interest is stationary (i.e., if the structure of the problem, the rules of the game, or the parameters of the problem stay the same for all stages), then any stationary (i.e., stable) set of decisions applied to a large enough number of decision stages will lead to either recurrent visits to one state or a systematic pattern of visits to a sequence of states. This is an abstraction from reality, to be sure, since the world itself is always changing, but this abstraction allows us to wrap our heads around problems that would otherwise be intractable.

This abstraction leads us to the notion of a steady state, which we encountered in Chap. 10. A **steady-state equilibrium probability distribution** is the set of transition probabilities that characterizes the long-run behavior of a Markov chain. Although wordy, it only means that there are underlying probabilities that correspond to the states most likely to be observed if you watch a system for long enough. Even if we know a system is unlikely to be always steady or the same, we can be reasonably confident of predicting outcomes, if we rely on long-run probabilities. We do this all the time when making long-term weather forecasts.[5]

---

[5] For most of modern history, human observers tended to think of the short-run fluctuations of rainfall and temperature (i.e., weather) as variable but the underlying climate generating those

Those who play *Monopoly* repeatedly begin to form a sense of which properties are most likely to be landed on by players, and therefore which properties might be good ones to purchase given the chance. Without being aware of it, players are actually solving for the long-run, steady-state equilibrium probability distribution for the *Monopoly* Markov chain.[6]

A theorem due to Isaacson and Madsen (1976) states that for any s-state ergodic chain there exists a vector $\pi = [\pi_1\ \pi_2\ \ldots\ \pi_s]$ such that:

$$\lim_{n \to \infty} \mathbf{P}^n = \begin{bmatrix} \pi_1 & \pi_2 & \cdots & \pi_s \\ \pi_1 & \pi_2 & \cdots & \pi_s \\ \vdots & \vdots & \ddots & \vdots \\ \pi_1 & \pi_2 & \cdots & \pi_s \end{bmatrix} \tag{12.3}$$

In other words, as the number of stages increases, the entries in the transition probability matrix begin to converge on their long-run values and the matrix develops identical rows. Furthermore, regardless of the initial state, the probability of being in state j is simply π_j. This provides an easy method for determining the long-run probabilities because, in the long run, the following relationship must hold:

$$\pi = \pi \mathbf{P}. \tag{12.4}$$

This property is best illustrated through an example.

Example: An Ergodic Markov Chain for Soft Drink Marketing

This example is based on a scenario posed by Winston (1991). Consider an industry consisting of two soft drink manufacturers. Suppose the entire industry produces only two colas. Each manufacturer produces one cola which competes head-to-head with the other. Imagine further that cola drinkers are loyal to their brand, but not fully so. Specifically, given that a consumer last purchased cola 1, there is a 90% chance that their next purchase will be cola 1, but a 10% chance that they will switch to cola 2. Similarly, given that a person last purchased cola 2, there is an 80% chance that their next purchase will be cola 2, but a 20% chance they will switch brands. The associated transition probability matrix is:

---

fluctuations as relatively stable and best modeled as a stationary system. We now know that, because of greenhouse gas forcing, the climate is changing and is therefore better thought of as a non-stationary system, which raises important hurdles for modeling climate-driven events such as drought or flooding.

[6] Dedicated *Monopoly* players will want to consult Ash and Bishop (1972) who derive the steady-state probability of occupying each square on the *Monopoly* board.

$$Q = \begin{bmatrix} 0.90 & 0.10 \\ 0.20 & 0.80 \end{bmatrix}.$$

This transition probability matrix is ergodic, according to the definition above. So, we can ask of it the following question: what is the probability that a cola 2 purchaser will be purchasing cola 1 two purchases from now? The probability matrix after two periods is given by the product of the transition probability matrix:

$$Q = \begin{bmatrix} 0.90 & 0.10 \\ 0.20 & 0.80 \end{bmatrix} \begin{bmatrix} 0.90 & 0.10 \\ 0.20 & 0.80 \end{bmatrix} = \begin{bmatrix} 0.83 & 0.17 \\ 0.34 & 0.66 \end{bmatrix}.$$

In other words, the probability that a cola 2 purchaser will be purchasing cola 1 after two periods is prob $(x_2 = 1 | x_0 = 2) = Q_{21}(2)$ i.e., element $\{2,1\}$ of Q^2. That probability is 0.34.

The logic that underlies this process is an underpinning of infinite-stage dynamic optimization and we can expand this method to find the long-run probabilities associated with purchases of cola 1 and cola 2.

To find the "stable" long-run probabilities we can rewrite our equation as:

$$[\pi_1 \quad \pi_2] = [\pi_1 \quad \pi_2] \begin{bmatrix} 0.90 & 0.10 \\ 0.20 & 0.80 \end{bmatrix}$$

or

$$\pi_1 = 0.90\pi_1 + 0.20\pi_2$$
$$\pi_2 = 0.10\pi_1 + 0.80\pi_2.$$

Now it is only necessary to solve these equations for π_1 and π_2. Unfortunately, this pair of equations has an infinite number of solutions! However, it turns out that we can replace the second equation by the condition $\pi_1 + \pi_2 = 1$, since we require the probabilities to sum to one. Solving this new system:

$$\pi_1 = 0.90\pi_1 + 0.20\pi_2$$
$$1 = \pi_1 + \pi_2$$

results in the solutions $\pi_1 = 2/3$ and $\pi_2 = 1/3$. This means that, in the long run, there is a 2/3 chance that a consumer will purchase cola 1 and a 1/3 chance that they will purchase cola 2. This is true regardless of whether the individual started out as a cola 1 or cola 2 drinker. Of what value is this information regarding the long run probabilities? How might we use it? Let us continue with the example.

Suppose each of ten million cola customers makes one purchase per week and suppose each can of cola earns the manufacturer $0.25 in profit. Now imagine an advertising agency approaches the manufacturer of cola 1 with a money-back guaranteed advertising campaign. The contract states that the advertising agency will decrease the share of customers who defect and switch from cola 1 to cola 2 from 10% to 5%. In other words, the advertising campaign will strengthen brand loyalty for cola 1. Suppose the advertising campaign costs $5,000,000 per year. What should the company do?

Fortunately, the company's accountants are well-trained and familiar with ergodic Markov process. They know that, at present, their best guess is that, in the long run, 2/3 of all cola purchases will be for cola 1. Each purchase yields $0.25 profit, so total annual profits are currently:

$$0.66 \times 0.25 \times 52 \times 10,000,000 = \$80,666,666.$$

The advertising firm guarantees that the campaign they are offering will change the transition probability matrix to:

$$\mathbf{Q} = \begin{bmatrix} 0.95 & 0.05 \\ 0.20 & 0.80 \end{bmatrix}.$$

Note that the firm guarantees a change in behavior for cola 1 purchasers, but no change in behavior for cola 2 purchasers (and hence any cola 1 purchasers who switch brands). If we solve for the steady-state transition probabilities following the procedure outlined above, we obtain the solution values $\pi_1 = 0.80$ and $\pi_2 = 0.20$ This means the company's new profits (under the advertising contract) would be:

$$0.80 \times 0.25 \times 52 \times 10,000,000 = \$100,400,000.$$

The gain in annual profits ($100,400,000 − 80,666,666 = $10,733,334) exceeds the annual cost of the contract ($5,000,000), so it makes sense for the company to hire the firm.

Example: Valuing a Sports Team

The Markovian approach can be easily applied to human resources problems such as modeling a workforce consisting of multiple employees with different career trajectories. This example, which is based on a problem originally posed by Flamholtz et al. (1984), is interesting because it incorporates an absorbing state. Suppose a sports team (pick your favorite) consists of 2 highly-paid stars, 13 reasonably-paid starters, and 10 substitutes. For tax purposes, the team owner must value the team, which consists of the total value of all players. The value of each player is defined to be the total value of the salary he will earn until retirement. How might we value the team?

Let's begin with a rather naïve sketch of the problem. Let's imagine that, at the beginning of each season, the players can be classified into one of four categories, as identified in Table 12.1.

Table 12.1 Hypothetical team members and their salaries

Category	Type	Salary ($/year)	Number
1	Star	5,000,000	2
2	Starter	1,000,000	13
3	Substitute	250,000	10
4	Retired	0	–

We could simply assume that those players who fall into the first three categories have a potentially active career of, say, 5 years, before they slip into retirement. Let's also assume that salary contracts are set at the start and cannot be renegotiated and that retirement is an absorbing state, i.e., retirees never come out of retirement. One approach would be to calculate the value of the team based on salaries and expected careers. Without discounting, that sum is simply:

$$\sum_{1}^{5}(5,000,000 \times 2 + 1,000,000 \times 13 + 250,000 \times 10) = \$127,500,000$$

If we instead wanted to discount the value of the team, then the discounted sum would be calculated as:

$$\sum_{1}^{5}\frac{1}{(1+\delta)^t}(5,000,000 \times 2 + 1,000,000 \times 13 + 250,000 \times 10)$$

which with $\delta = 5\%$, equals a present discounted value of \$110,401,655.

More realistically, due to injuries and variable performance, we might suppose that players on a team move in and out of the categories listed above. For example, given that a player is a star, starter, or substitute at the beginning of the current season, we might assign probabilities that they will become a star, a starter, a substitute, or retired at the beginning of the next season. Consider the transition probability matrix contained in Table 12.2.

In Table 12.2 rows correspond to the status in the current season and columns correspond to next season. So, for example, stars only have a 50–50 chance of remaining stars, and substitutes only have a 1% chance of becoming stars. Note from the final row that we still assume retirement is an absorbing state: a player

Table 12.2 Transition probabilities for categories of players

This year	Next year			
	Star	Starter	Substitute	Retired
Star	0.50	0.33	0.16	0.01
Starter	0.20	0.60	0.15	0.05
Substitute	0.01	0.09	0.80	0.10
Retired	0.00	0.00	0.00	1.00

who is retired this year has zero probability of returning as a star, starter or substitute next year.

We know that eventually all players will enter retirement and the owner will no longer need to pay them. (From a practical point of view, let's say the owner can hire new players in the background of this problem to keep the team intact and viable against opponents.) How long will it take for all players to enter the absorbing state? It turns out that the expected steps to absorption can be found via the formula:

$$N = (I - Q)^{-1}$$

where I is the identity matrix, Q is the transient state submatrix (i.e., the 3×3 matrix for stars, starters and substitutes) and N is the column vector with the expected number of steps for each category of player. In this case, the expected times to retirement are:

Stars:	~ 6.6 years
Starters:	~ 5.1 years
Substitutes:	~ 3.3 years.

Rounding these to the more meaningful integer values 7, 5, and 3, we can generate a more realistic profile of the value of the team, using $\delta = 5\%$ and the projected expected length of career for each category of player.

Over the entirety of their time with the team, each star can be expected to receive (in net present value terms):

$$\sum_{1}^{7} \frac{5,000,000}{(1+0.05)^t} = \$28,931,867.$$

Each starter is expected to receive:

$$\sum_{1}^{5} \frac{1,000,000}{(1+0.05)^t} = \$4,329,477.$$

Each substitute is expected to receive:

$$\sum_{1}^{3} \frac{250,000}{(1+0.05)^t} = \$680,812.$$

With two stars, 13 starters and 10 substitutes, and taking into account shifts between categories during a player's career, we arrive at a more accurate total expected net present value for the team, which is \$120,955,055.

References

Ash, R., & Bishop, R. (1972). Monopoly as a Markov process. *Mathematics Magazine, 45*, 26–29.

Flamholtz, E., Geis, G., & Perle, R. (1984). A Markovian model for the valuation of human assets acquired by an organizational purchase. *Interfaces, 17*, 11–15.

George, H. (1881). *Progress and poverty*. D. Appleton and Company.

Isaacson, D., & Madsen, R. (1976). *Markov chains: Theory and applications*. Wiley.

Leopold, T. 2015. Monopoly: At 80, it just keeps Go-ing. *CNN March 19, 2015*. Retrieved April 08, 2025, from https://www.cnn.com/2015/03/19/living/feat-monopoly-80th-anniversary

Stewart, I. (1996, October). Mathematical recreations: Monopoly revisited. *Scientific American*, 116–119.

Winston, W. L. (1991). *Operations research: Applications and algorithms*. PWS-Kent Co.

To Infinity ... and Beyond

13

As the examples from Chap. 12 illustrate, when it comes to problems that are both probabilistic and intertemporal, what you initially see isn't always what you eventually get. When there are transitions between states, characterizing the problem in terms of a Markov process allows us to project into the future and think about what the system might look like when outcomes begin to converge toward their steady-state values. It turns out that characterizing some problems as never ending has some advantages.

The assumption of an infinite horizon is of course a mathematical convenience since planning horizons are really never truly infinite. But formulating a problem that has a large but finite number of stages as an infinite-horizon problem is often useful in so far as the infinite horizon is a reasonable approximation to a long but finite horizon. Although infinite-horizon problems may be more difficult to solve than finite-horizon problems, the analysis of infinite-horizon problems can be a bit more elegant.

The main assumption that we need to maintain when setting-up and solving an infinite-horizon problem is stationarity. As we've covered before, stationarity implies that the equations and statistics that determine values in the problem do not change across stages. Note, however, that this does not require *values* to stay the same, only that the processes which determine those values stay the same. This is because infinite-horizon problems are concerned with strategies and solutions that are stable in the long run, i.e., strategies that perform the same whenever we encounter the same state conditions, regardless of the stage of the problem. In other words, the optimal decision vector for stationary problems with infinite decision stages will be invariant to the stage (e.g., time period) of the problem. This, of course, does not mean that a single equilibrium or decision is optimal for all states. It means that the optimal decision should depend *only* on the current state and not on the decision stage.

G. Shively, *A Beginner's Guide to Dynamic Optimization in Economics*, Classroom Companion: Economics,
https://doi.org/10.1007/978-3-032-09374-5_13

Bellman's Equation Revisited

The easiest kind of infinite-horizon dynamic programming problem to solve is one that includes discounting. This is because a steady-state solution eventually returns a constant stream of finite returns and the present value of the infinite stream of stage returns will converge to a finite sum for any positive discount rate. For this reason, when an infinite number of stages remain in a problem the value of each state at the i–th decision stage $V_i\{x_i\}$ is finite and equal to $V_{i+1}\{x_{i+1}\}$.

As a result, Bellman's equation for the infinite-stage, stationary problem with discounting is the same as Bellman's equation for the finite-stage DP problem with the stage subscripts removed:

$$V\{x\} = \ max \ [F(x, u) + \ \beta \ V\{r(x, u)\}]. \tag{13.1}$$

The most important feature to note about Eq. 13.1 is that $V\{x\}$ appears on both sides of the equation. In the long run, the optimal value function is said to "confirm itself." This means that if an optimal policy is being followed, the best one can hope to achieve is a return to exactly the same position in the subsequent period. This feature leads to two different approaches to solving a DP problem: **value iteration** and **policy iteration**. Both are forms of **successive approximation** in which one starts with an initial guess at a solution and then refines the guess until one is satisfied that an optimal solution has been reached. In the former, one starts with an arbitrary value for the value function $V\{x\}$ and attempts to find better values, iterating until no improvement in the value function can be made. Whatever policy gave rise to the optimum is then deemed the optimal policy. Under the latter approach, one iterates across available policies, calculating the value function for each possible state under that policy. Once a stable policy is found the process ends. In general, value iteration tends to work well and quickly for problems with a small state space; policy iteration can converge on a solution more quickly but can be computationally demanding.

To see how this works in practice we use the example of harvesting trees from a forest planted for commercial purposes. We start by taking a few steps backward to set up the problem in its simplest form before complicating it.

Harvesting Timber

Consider a world without uncertainty in which an investor owns a plot of land on which trees are growing. Imagine the trees are of similar age and type. Such plantation forests (what foresters call an even-aged stand) differ little from a field of corn except that it takes trees longer to grow to maturity—often many decades depending on the species. Among trees, Loblolly pine (*Pinus taeda*) is one of the fastest growing, which makes it particularly attractive as a commercial species. It is grown throughout the Southeastern United States both for lumber and for paper

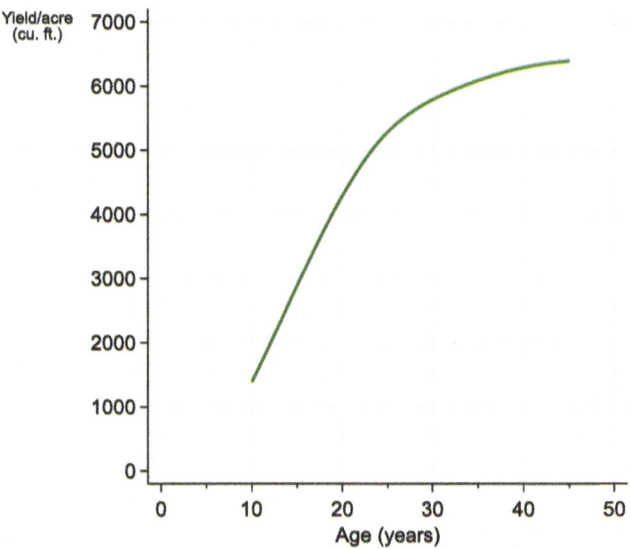

Fig. 13.1 Biomass accumulation of Loblolly pine

pulp. Figure 13.1, which is derived from a U.S. Forest Service field study by Smalley and Bailey (1974) shows that between establishment and year 40, Loblolly pine grows rapidly. Then, starting around age 50, growth begins to level off and no further appreciable accumulation of merchantable biomass occurs.[1]

Looking at Fig. 13.1, one might ask: at what age should the forest owner harvest the trees? It turns out that there are at least three possible answers to this question.[2] Let's consider them each in turn. For simplicity, let's imagine that the price of timber net of the cost of harvesting and marketing trees remains constant over time (in real terms) and the health of the trees is assured.

In the first case, let's call it Scenario 1, one might speculate that the landowner wants to maximize biomass (yield). In this case, the "optimal" decision will be to let the trees grow until the accumulation of biomass is at a maximum, which in Fig. 13.1 would likely occur around year 50. The word "optimum" here is placed in

---

[1] Note that Fig. 13.1 doesn't include data for years 0–10 or years beyond 45 because Smalley and Bailey only started measuring trees once they reached 10 years of age and didn't observe stands much older than 45 years. Harvesting outside the years observed was extremely unlikely in their commercial sample.

[2] If one includes additional considerations, such as the **ecosystem services** trees provide (e.g., wildlife habitat, watershed protection, or the carbon they sequester), then there will certainly be more potential answers. Here we keep the focus narrowly on the economic value of the harvested trees. However, the idea that ecosystem services generate value which should be paid for has gotten traction in a number of settings. For example, in 2003 Mexico created *Pago por Servicios Ambientales Hidrológicos* (the Payment of Hydrological Environmental Services Programme), which pays for the conservation of forests in hydrologically-critical watersheds using revenue from charges levied on downstream water users.

quotation marks because while this approach maximizes yield, which may be
desirable from a biomass perspective, it is decidedly sub-optimal from an economic
perspective.

To see why, let's consider Scenario 2. Imagine that the landowner, as an economic
agent, wants to maximize her return on invested capital. At any point, she can clear the
stand, sell the trees and place the money in a safe bank account that returns
a guaranteed rate of interest. In the early years, when trees are growing rapidly, it
will make economic sense for our investor to leave her capital growing in the form of
trees—that is, unless interest rates are extremely high. After all, when trees are young,
they are adding value faster than savings in the bank would. In later years, when trees
are old and not growing so rapidly, it makes sense for our investor to take her capital
out of the trees and put the proceeds in the bank—in this case, unless interest rates are
extremely low. This is because old trees are not adding value very fast in comparison
to what savings would generate in the bank.[3] The economic logic in this scenario is
that our forest owner will be indifferent between harvesting and not harvesting exactly
at the point where the growth rate of trees matches the interest rate. Figure 13.2
illustrates where this point occurs in the case of $r = 5\%$. In this case, the economically
optimal maturity for a single rotation is approximately 25 years, with a yield of

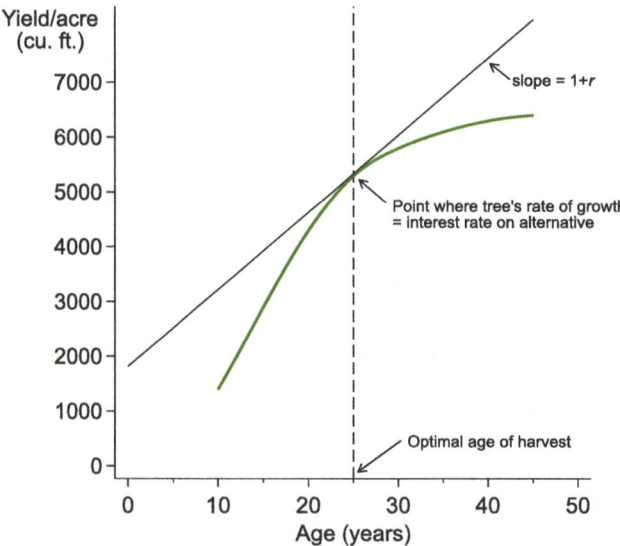

Fig. 13.2 Matching the rate of tree growth to the interest rate

---

[3] It seems necessary to again emphasize that we are ignoring the important ecosystem services
that old and dying trees might generate. These services include nesting areas for birds and
animals, and decaying material to feed fungi. To paraphrase Udry (1956) a tree, even an old
one, is nice.

approximately 5,300 cubic feet of lumber per acre. Prior to this point, the curve tracking the growth of trees is steeper than the interest rate line. Beyond this point, the growth rate of trees is less than the interest rate. Beyond year 25, leaving the trees to grow will continue to add economic value but not as rapidly as savings in a bank paying 5% on deposits. It should be obvious that we could pivot the line corresponding to the return on savings in the bank: a steeper slope is tangent to the growth curve at a lower point, encouraging earlier cutting. A lower interest rate is tangent to the growth curve at a higher and flatter point, which suggests later cutting.

Although the problem of choosing when to harvest trees in Scenarios 1 and 2 involves time, it is not, strictly speaking, a dynamic problem. The reason is that we've set up the problem in such a way that once trees are harvested, our landowner simply walks away and heads to the bank. There is no sense in which future decisions depend on this cutting decision. But recall that the primary feature of dynamic optimization, which was introduced all the way back in Chap. 1, is that current decisions have implications for future decisions. Let's stop and reflect on this for just a moment. If we think of the state variable for this problem as a description of trees on the land (trees of a certain age vs. no trees) and replanting can occur following harvest, then the problem becomes a bit more interesting and considerably more dynamic. What happens to the land in question after the trees are harvested? Does it just remain fallow or is it put to use again? If it is possible to plant a new stand of trees and begin the process anew, then might this influence our decision about when to harvest? This leads us to contemplate Scenario 3.

In terms of forest management, we are now stepping into what foresters refer to as a **rotation problem**, i.e., determining when to cut and re-plant to maximize the value of forest land, not just the value of the trees growing on the land. To make this clear, let's formalize things a bit and call $V(T)$ the value of a rotation of length T. In Scenario 2, we determined that $V(25) \approx 5,300$. But now let's consider a problem in which multiple rotations are possible. For example, Fig. 13.3 illustrates a series of three possible rotations, each 25 years in length.

The return to this series of rotations, call it V, is simply:

$$V = \left(\frac{1}{1+r}\right)^{25} V(25) + \left(\frac{1}{1+r}\right)^{50} V(50) + \left(\frac{1}{1+r}\right)^{75} V(75)$$

or

$$V = \left(\frac{1}{1.05}\right)^{25} 5,300 + \left(\frac{1}{1.05}\right)^{50} 5,300 + \left(\frac{1}{1.05}\right)^{75} 5,300$$

which is

$$1,565 + 462 + 136 = 2,163.$$

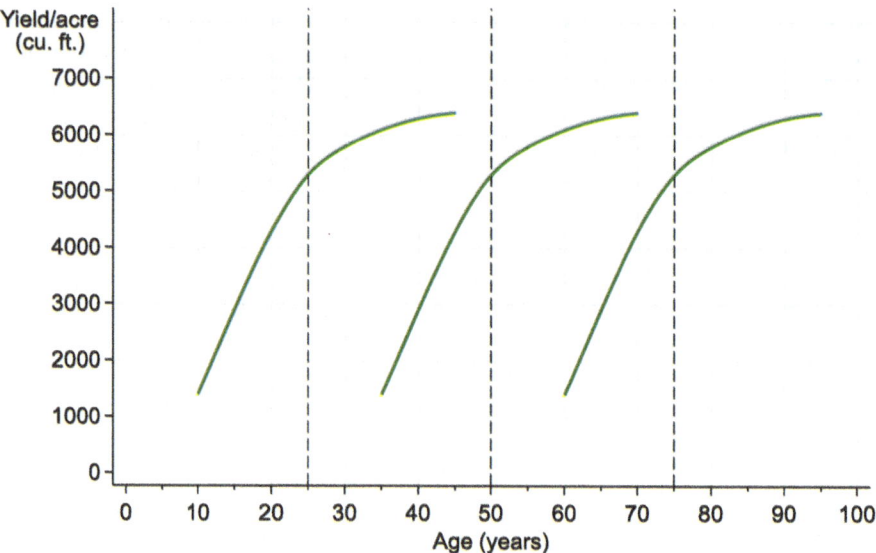

Fig. 13.3 Three tree rotations: plant, harvest, repeat

With three possible rotations, is 25 years now the optimal rotation length? What if we can plant trees in perpetuity? In other words, what if the land is valued on the basis of an infinite series of rotations?

The Faustmann Model

In the mid-nineteenth century, a German forester named Martin Faustmann stumbled upon this problem and proposed a solution.[4] The **Faustmann model** characterizes the optimal timing of forest harvests to maximize the present value of net revenues over an infinite time horizon. The Faustmann formula calculates what is known as the **Land Expectation Value (LEV)** or **Soil Expectation Value (SEV)**, which represents the present value of a bare piece of land to be used for an infinite series of timber rotations. In its classic form, the Faustmann model is completely deterministic: all biological and economic parameters are known and fixed for all time.

Although Faustmann didn't have the twentieth century tools of the Bellman equation available to him in 1849, his thinking proceeded as if he did. The standard Faustmann model setup defines $V(t)$ as the value of the tree stand at age t; $R(t)$ as the revenue from harvesting at age t; C as the cost of planting a new stand of trees,

---

[4] Faustmann, who was born in 1822, proposed his forest rotation model in 1849. It was largely ignored for a century but is now considered the benchmark model for determining optimal forest rotations. It is central to natural resource economics and continues to be adapted and applied to a range of forestry problems. For a review of the model and its impacts, see the symposium papers summarized by Brazee (2001).

and r as the interest rate. The corresponding Bellman equation expresses the potential value of land as consisting of two pieces: (i) revenue minus cost for the rotation length under consideration; and (ii) the discounted value of the future stream of earnings if the rotation is extended by a small increment of time, Δt:

$$V_t = max\{R_t - C, e^{-rt}V_{t+\Delta t}\}. \tag{13.2}$$

In the Faustmann model, because the decision to be made is whether to harvest now or to wait, and nothing is uncertain, the value function is typically written as the present value of a perpetual series of rotations:

$$V = \max_{T}\left\{\frac{R_t - C}{1 - e^{-rT}}\right\} \tag{13.3}$$

where T, the rotation age, is the decision variable and the denominator $1 - e^{-rT}$ accounts for the infinite repetition of rotations every T years. What the **Faustmann formula** tells us is that when confronted with the possibility of replanting in perpetuity, the standard single-period rotation length is not optimal. Instead, the optimal rotation will be slightly shorter. The reason for this is that shortening the initial rotation (and all future rotations) *slightly* means that the infinite stream of returns from future harvests are moved a bit closer to the present. Admittedly, these future sums are small, especially with discounting, but moving them just a bit closer to the present increases the value function. Of course, there is a limit to how much overall benefit can be achieved by shortening the initial and subsequent rotations because shortening the length of the rotation requires one to give up some growth in biomass at each harvest. As a result, finding the optimal rotation requires balancing this loss against accelerating the arrival of future gains.

If, for example, we return to our series of three rotations, but we shorten them slightly, say by one year, yield falls slightly and the corresponding value function changes as follows:

$$V = \left(\frac{1}{1+r}\right)^{24}V(24) + \left(\frac{1}{1+r}\right)^{48}V(48) + \left(\frac{1}{1+r}\right)^{72}V(72)$$

or

$$V = \left(\frac{1}{1.05}\right)^{24}5,100 + \left(\frac{1}{1.05}\right)^{48}5,100 + \left(\frac{1}{1.05}\right)^{72}5,100$$

where the power to which we are discounting declines due to the shortened rotation length. The sum is:

$$1,581 + 490 + 152 = 2,223$$

which is slightly larger than what we found previously. Although the first harvest value is smaller than in the prior case, because we've forfeited some growth, the second and third terms are slightly larger than before because they have been moved closer and hence are less diminished by discounting. If we were to extend the analysis to an infinite horizon, this basic principle would continue to exert influence over the decision and nudge T shorter still. A higher discount rate de-emphasizes future harvests and favors the first rotation. In contrast, a lower discount rate favors future harvests over the first rotation.

How might value iteration and policy iteration apply to the rotation problem? The example we've laid out is trivial, largely because the state space includes no uncertainty. We know exactly how much yield is available at year 30 if we postpone cutting in year 29. And while the Faustmann formula, specifically Eq. 13.3, provides an exact solution for the optimal rotation length, we could also find it through a simple process of trial and error. For example, we could try successive policies (e.g., cut in year 20, year 21, year 22, etc.) and compute the values of the value function associated with each policy. The optimal policy, then, would be the one that provided the largest value for the infinite sum. Or, reading down a list of values, we could look for the largest one and then figure out which policy gave rise to it. In both cases, we would be using crude forms of successive approximation—policy iteration and value iteration—to find our solution. At the end of this chapter, we will work through a more complicated example.

Before doing so, however, it is important to point out that Faustmann's original formula, strictly speaking, is a special case of a Markov decision process in which the transition probabilities between states (i.e., between growth categories for the trees) are either ones or zeros. Imagine, if you will, a transition probability matrix for tree growth. If biomass at year 20 follows directly and deterministically from biomass at year 19, the corresponding entry in the transition probability matrix will be 1. No other state (i.e., biomass) is possible because there is nothing unpredictable about tree growth in our deterministic example. However, like any other biological system, a forest is subject to random shocks and events: a storm might topple some trees, a fire might sweep through the stand, or a pest infestation might reduce tree growth. Some trees may grow a little faster because they get just a bit more sunlight or happen to be receiving slightly better nutrients, and some may grow a little slower for opposite reasons. Similarly, some of the underlying economic variables, such as the price of timber or the interest rate, may be uncertain. As a consequence, a stochastic version of the rotation problem is probably more realistic. In this case, there is not a one-to-one matching between rotation age and stand biomass. Instead, there may be a probabilistic transition between adjacent states. For example, a 15-year-old one-acre stand of trees consisting of 3,000 cubic feet of merchantable timber might in year 16 consist of 2,900 cubic feet (because some trees were blown over in a wind storm) or it might contain 3,100 cubic feet, or 3,200 cubic feet, or some other amount, all with some probability of occurrence. Similarly, the prices of forest products and costs of production may vary such that they may be high or low, again with some assigned probabilities.

As we add this kind of stochastic complexity to the problem, the **state-space** begins to expand. Consider the possibility of three growth outcomes in year 15, and three potential prices. This implies nine different possible states, all with different associated probabilities, all of which must be examined when considering potential optimality of a rotation strategy. With three potential growth outcomes, three different prices, and three different costs, our state-space expands to 27 possibilities. With five potential outcomes for each variable, we end up with 125 different states to consider. This tendency for a problem to explode in size as state variables and potential states are added is referred to as the **curse of dimensionality**. This curse means that making problems more realistic, in terms of mimicking the number and uncertainty of actual states of the world, tends to make them more intractable, in terms of computational burden.

Building on Faustmann, Buongiorno (2001) outlines and discusses in detail a stochastic version of Faustmann's formula, which is:

$$\max_{d_t} \; V(X_0) = E\left\{ \sum_{t=0}^{\infty} \left(\frac{1}{1+r} \right)^t R(X_t, d_t) | X_0 \right\} \tag{13.4}$$

where d_t is the decision taken at time t, r is the interest rate, X_t describes the system at time t (which is now a random variable), and $R(X_t, d_t)|X_0$ denotes the return realized at time t when decision d_t is taken in state X_t, which is understood to potentially depend on the initial conditions X_0. Considering growth only, the expectation operator in Eq. 13.4 replaces the deterministic growth function with a set of transition probabilities between growth states and their associated values. The objective function becomes the expected discounted value of returns over an infinite horizon, accounting for the probabilistic nature of future events.

Policy Iteration

We close this chapter by working through an infinite-horizon stochastic dynamic programming problem using policy iteration. The problem, which is inspired by Sells (1995), follows the example and treatment of Bertsekas (1987; see Chap. 5). It is highly stylized and simple in its set up but clearly demonstrates the mechanics behind the iterative approach to solving a dynamic programming problem when states are probabilistic and the horizon is infinite. As its name implies, policy iteration uses an iterative, three-step procedure which is guaranteed to converge in a finite number of iterations. The three steps are as follows:

Step 1:   Choose any initial policy to follow;
Step 2:   Calculate the corresponding benefit (or, as in this case, cost) function;
Step 3:   Obtain an improved policy, using information about the optimal cost associated with movements in a one-stage version of the problem.

With policy iteration, at each iteration a system of equations must be solved. The size of the system depends on the number of states (one equation per state).

Consider a farmer who is battling invasive weeds, which is a perennial problem for anyone who has ever tried to raise a garden. In each season, the farmer faces two possible states, call them:

$$x = \{1, 2\}$$

where $x = 1$ indicates low weed emergence and $x = 2$ indicates high weed emergence. The farmer has available one of two herbicides. His choice is represented by the decision vector:

$$d = \{d^1, d^2\}$$

here d^1 is a highly-effective herbicide, which is costly, and d^2 is a lower-cost alternative, which is less effective.

We assume transition probabilities follow an ergodic Markov process with the following transition probability matrices:

$$\rho(d^1) = \begin{bmatrix} \rho_{11}(d^1) & \rho_{12}(d^1) \\ \rho_{21}(d^1) & \rho_{22}(d^1) \end{bmatrix} = \begin{bmatrix} 0.80 & 0.20 \\ 0.80 & 0.20 \end{bmatrix}$$

$$\rho(d^2) = \begin{bmatrix} \rho_{11}(d^2) & \rho_{12}(d^2) \\ \rho_{21}(d^2) & \rho_{22}(d^2) \end{bmatrix} = \begin{bmatrix} 0.20 & 0.80 \\ 0.20 & 0.80 \end{bmatrix}.$$

We denote decisions by the superscripts and the associated probabilities by subscripts. Each entry in the transition probability matrix indicates the probability of moving between states. There's a lot of notation here, with lots of subscripts and superscripts floating around, so let's break it down.

First note that there are two transition probability matrices, one for each possible decision, these are $\rho(d^1)$ and $\rho(d^2)$. You can think of each decision as resulting in a different set of potential transitions between situations. These transitions are represented in the matrices by the entries $\rho_{ij}(d^k)$, where i indexes the current state (low or high weed emergence) and j indexes the subsequent state (low or high weed emergence), conditional on decision k. Each decision (i.e., high cost, high effectiveness vs. low cost, low effectiveness) implies a cost and also implies probabilities of weed emergence. Although nothing is guaranteed, since this is a stochastic problem, when the farmer purchases the high-cost herbicide he is in essence purchasing a different probability distribution, just as the cola company in Chap. 12 was buying from the advertising agency a different set of transition probabilities among consumers.

Picking a probability out of the first matrix, say $\rho_{21}(d^1)$ we can see that the probability of moving from a state of "high weeds" ($i = 2$) to a state of "low weeds" ($j = 1$) using a high-effectiveness herbicide (d^1) is $\rho_{21}(d^1) = 0.80$. In other words, the high-cost herbicide is effective, but not completely so, since there remains some probability of high weed emergence. Similarly, entry $p_{11}(d^2)$ from the second matrix tells us that the probability of moving from a state of "low weeds" ($i = 1$) to a state of "low weeds" ($j = 1$) using a low-effectiveness herbicide (d^2) is $p_{11}(d^2) = 0.20$. In other words, the low-cost herbicide leaves the field vulnerable to high weed emergence.

Let us now define **transition costs** associated with each decision as follows:

$$c(1, d^1) = 1.80$$
$$c(1, d^2) = 0.25$$
$$c(2, d^1) = 1.00$$
$$c(2, d^2) = 2.50.$$

We call these transition costs because "cost" implies the "purchase" of a strategy (possibly including the damage to crops associated with that strategy). In other words, we purchase a potential transition state as we enter the next stage of the problem. For example, the cost function says that if state 1 prevails (weed emergence is low) and the high-cost strategy d^1 is used (purchased), the cost is $c(1, d^1) = 1.80$. Our farmer can strategically purchase a transition probability and the price of obtaining this probability is the cost of the strategy.

Let's allow the farmer to discount, and set the per-period discount factor at $\beta = 0.95$. Can we determine the best policy to follow, in the sense that the optimal decision will only depend on the state and not how we reached that state or what stage we are in? That's our gold standard for a DP solution. It should depend only on the current state and not on the past. In other words, it should take the form "if I find myself in state x, always make decision d." This will be a stable strategy in the sense that while it might not always work out in the short run, in the long run it is the strategy that results in the lowest expected cost.

To characterize the basic form of the Bellman equation, let the state $= i$ and the decision (control) equal d. We seek to minimize the **cost-to-go** function:

$$T(V)(i) = min \left\{ \begin{array}{c} c(i, d^1) + \beta \sum_{j-1}^{2} p_{ij}(d^1) V(j), \\ c(i, d^2) + \beta \sum_{j-1}^{2} p_{ij}(d^2) V(j) \end{array} \right\}. \tag{13.5}$$

In Eq. 13.5, $T(V)(i)$ is shorthand notation for a mapping which represents the optimal cost for a single-stage problem given initial state i, stage cost $c(i,d)$ and the subsequent value function in the next stage of the problem, $V(j)$. The solution method follows three iterative steps, as follows:

Step 1:   Start with any initial, state-contingent policy for decisions at stage 0. We can choose, for example, to implement decision 1 (high-cost herbicide) for both states, i.e., $d^i(1) = d^1$, $d^i(2) = d^2$. It doesn't matter what set of state-contingent policies we begin with.

Step 2:   Obtain the initial value function V^0 (with the superscript 0 indicating the initial stage) as:

$$V^0(1) = c(1, d^1) + \beta[\rho_{11}(d^1) \times V^0(1) + \rho_{12}(d^1) \times V^0(2)]$$

$$V^0(2) = c(2, d^1) + \beta[\rho_{21}(d^1) \times V^0(1) + \rho_{22}(d^1) \times V^0(2)].$$

Substituting the data from the problem:

$$V^0(1) = 1.80 + 0.95[0.80 \times V^0(1) + 0.20 \times V^0(2)]$$

$$V^0(2) = 1.00 + 0.95[0.80 \times V^0(1) + 0.20 \times V^0(2)]$$

Solving these two equations for the two unknowns gives us:

$$V^0(1) \cong 32.96 \text{ and } V^0(2) \cong 32.16.$$

Step 3:   To find a potential policy improvement, substitute these values into V and find the policies that minimize cost in each state:

$$\text{state 1}: \mathrm{T}(V^0)(1) = \min\{1.80 + 0.95[0.80 \times 32.96 + 0.20 \times 32.16],$$
$$0.25 + 0.95[0.20 \times 32.96 + 0.80 \times 32.16]\}$$
$$= \min\{32.96, 30.95\} = 30.95$$

$$\text{state 2}: \mathrm{T}(V^0)(2) = \min\{1.00 + 0.95(0.80 \times 32.96 + 0.20 \times 32.16),$$
$$2.50 + 0.95(0.20 \times 32.96 + 0.80 \times 32.16)\}$$
$$= \min\{32.16, 33.20\} = 32.16$$

This tells us the minimizing controls are d^2 for state 1 and d^1 for state 2. Now obtain the stage 1 value function V^1 using $V^1 = T^1(V^1)$ as in step 1:

$$V^1(1) = c(1, d^2) + \beta[\rho_{11}(d^2) \times V^1(1) + \rho_{12}(d^2) \times V^1(2)]$$

and

$$V^1(2) = c(2, d^1) + \beta[\rho_{21}(d^1) \times V^1(1) + \rho_{22}(d^1) \times V^1(2)].$$

Substituting data gives:

$$V^1(1) = 0.25 + 0.95[0.20 \times V^1(1) + 0.80 \times V^1(2)]$$

and

$$V^1(2) = 1.00 + 0.95\left[0.80 \times V^1(1) + 0.20 \times V^1(2)\right].$$

Solving yields $V^1(1) = 12.26$ and $V^1(2) = 12.73$.
Now iterate, and follow Step 3 again:

$$
\begin{aligned}
T(V^1)(1) &= \min\{1.80 + 0.95[0.80 \times 12.26 + 0.20 \times 12.73], \\
&\qquad\quad 0.25 + 0.95[0.20 \times 12.26 + 0.80 \times 12.73]\} \\
&= \min\{13.54, 12.26\} = 12.26
\end{aligned}
$$

which indicates convergence, and

$$
\begin{aligned}
T(V^1)(2) &= \min\{1.00 + 0.95[0.80 \times 12.26 + 0.20 \times 12.73], \\
&\qquad\quad 2.50 + 0.95[0.20 \times 12.26 + 0.80 \times 12.73]\} \\
&= \min\{12.73, 14.50\} = 12.73
\end{aligned}
$$

which also indicates convergence. We conclude, therefore, that the minimizing controls are d^2 in state 1 and d^1 in state 2. In other words, the farmer should use the low-cost herbicide when weed emergence is low and the high-cost herbicide when weed emergence is high.[5]

References

Bertsekas, D. P. (1987). *Dynamic programming: Deterministic and stochastic models.* Prentice-Hall, Inc.

Brazee, R. J. (2001). The Faustmann formula: Fundamental to forest economics 150 years after publication. *Forest Science, 47*(4), 441–442.

Buongiorno, J. (2001). Generalization of Faustmann's formula for stochastic forest growth and prices with Markov decision process models. *Forest Science, 47*(4), 466–474.

Sells, J. E. (1995). Optimising weed management using stochastic dynamic programming to take account of uncertain herbicide performance. *Agricultural Systems, 48*(3), 271–296.

Smalley, G. W., & Bailey, R. T. (1974). *Yield tables and stand structure for loblolly pine plantations in Tennessee, Alabama, and Georgia Highlands* (Forest Service Research Paper SO-96). U.S. Department of Agriculture Forest Service, Southern Forest Experiment Station.

Udry, J. M. (1956). *A tree is nice.* Harper and Brothers.

---

[5] Ideally, of course, we might hope that scientists could someday find viable alternatives to using any herbicides at all. Such a development is not beyond reach. For example, the agronomist and crop breeder Gebisa Ejeta won the World Food Prize in 2009 for his work in developing high-yielding varieties of sorghum, a staple cereal crop planted throughout Africa. These new varieties are naturally resistant to *Striga*, a parasitic weed that affects more than 100 million people across the African continent. Ejeta identified a gene that strengthens sorghum's resistance to *Striga*, thereby reducing the need for chemical methods of control.

The manufacturer's authorised representative in the EU is Springer
Nature Customer Service Centre GmbH, Europaplatz 3, 69115 Heidelberg,
Germany. If you have any concerns regarding our products, please
contact ProductSafety@springernature.com

Printed and bound by CPI Group (UK) Ltd, Croydon, CR0 4YY
23/04/2026
02095585-0017